KANT AND THE LAWS OF NATURE

Laws of nature play a central role in Kant's theoretical philosophy and are crucial to understanding his philosophy of science in particular. In this volume of new essays, the first systematic investigation of its kind, a distinguished team of scholars explores Kant's views on the laws of nature in the physical and life sciences. Their essays focus particularly on the laws of physics and biology and consider topics including the separation in Kant's treatment of the physical and life sciences, the relation between universal and empirical laws of nature, and the role of reason and the understanding in imposing order and lawful unity on nature. The volume will be of great interest to advanced students and scholars of Kant's philosophy of science and to historians and philosophers of science more generally.

MICHELA MASSIMI is Professor of Philosophy of Science at the University of Edinburgh. She has published widely on Kant and on the history and philosophy of science.

ANGELA BREITENBACH is Lecturer in Philosophy at the University of Cambridge and Fellow of King's College. Her publications focus on Kant, the philosophy of science, and aesthetics.

KANT AND THE LAWS OF NATURE

MICHELA MASSIMI
University of Edinburgh

ANGELA BREITENBACH
University of Cambridge

CAMBRIDGE
UNIVERSITY PRESS

CAMBRIDGE
UNIVERSITY PRESS

University Printing House, Cambridge CB2 8BS, United Kingdom

One Liberty Plaza, 20th Floor, New York, NY 10006, USA

477 Williamstown Road, Port Melbourne, VIC 3207, Australia

4843/24, 2nd Floor, Ansari Road, Daryaganj, Delhi – 110002, India

79 Anson Road, #06–04/06, Singapore 079906

Cambridge University Press is part of the University of Cambridge.

It furthers the University's mission by disseminating knowledge in the pursuit of education, learning, and research at the highest international levels of excellence.

www.cambridge.org
Information on this title: www.cambridge.org/9781107120983
10.1017/9781316389645

First published 2017

A catalogue record for this publication is available from the British Library.

Library of Congress Cataloging-in-Publication Data
NAMES: Massimi, Michela, editor.
TITLE: Kant and the laws of nature / [edited by] Michela Massimi, University of Edinburgh, Angela Breitenbach, University of Cambridge.
DESCRIPTION: New York : Cambridge University Press, 2016. |
Includes bibliographical references and index.
IDENTIFIERS: LCCN 2016040372 | ISBN 9781107120983 (Hardback : alk. paper)
SUBJECTS: LCSH: Kant, Immanuel, 1724-1804. | Science–Philosophy. | Natural law.
CLASSIFICATION: LCC Q175 .K1788 2016 | DDC 501–dc23 LC record available at https://lccn.loc.gov/2016040372

ISBN 978-1-107-12098-3 Hardback

Contents

Contributors

KARL AMERIKS is Hank McMahon Professor of Philosophy at the University of Notre Dame.

ANGELA BREITENBACH is University Lecturer in Philosophy and Fellow of King's College at the University of Cambridge.

MICHAEL FRIEDMAN is Suppes Professor of Philosophy of Science and Professor, by courtesy, of German Studies at Stanford University.

HANNAH GINSBORG is Professor of Philosophy at the University of California at Berkeley.

PAUL GUYER is Jonathan Nelson Professor of Humanities and Philosophy at Brown University.

MICHELA MASSIMI is Professor of Philosophy of Science at the University of Edinburgh.

JAMES MESSINA is Assistant Professor of Philosophy at the University of Wisconsin at Madison.

MARIUS STAN is Assistant Professor of Philosophy at Boston College.

THOMAS TEUFEL is Associate Professor of Philosophy at Baruch College, City University of New York.

DANIEL WARREN is Associate Professor of Philosophy at the University of California at Berkeley.

ERIC WATKINS is Professor of Philosophy at the University of California at San Diego.

CATHERINE WILSON is Anniversary Professor of Philosophy at the University of York and Distinguished Professor at the Graduate Center, City University of New York.

RACHEL ZUCKERT is Associate Professor of Philosophy at Northwestern University.

Note on Texts and Translations

All quotations from Kant's works refer to the volume and page numbers of the Akademie edition, that is, *Kants gesammelte Schriften* ed. Deutsche (formerly Königlich Preussische) Akademie der Wissenschaften, 29 vols. (Berlin: de Gruyter 1900–), except for quotations from the *Critique of Pure Reason*, which are referred to by A or B and the page number from the first or second edition of this work. Except where indicated otherwise, all translations use the *Cambridge Edition of the Works of Immanuel Kant*. Bibliographic information for all other works is supplied in the bibliography.

Abbreviations of Kant's Works

Anthr	Anthropology from a Pragmatic Point of View
CB	Conjectural Beginning of Human History
CF	Conflict of the Faculties
CJ	Critique of the Power of Judgment
Corr	Correspondence
CPR	Critique of Pure Reason
CPrR	Critique of Practical Reason
DS	Differentiation of Directions in Space
Disc	On a Discovery Whereby any New Critique of Pure Reason Is to Be Made Superfluous by an Older One
Dist	Inquiry Concerning the Distinctness of the Principles of Natural Theology and Morality
Dreams	Dreams of a Spirit-Seer
E	An Answer to the Question: What Is Enlightenment?
G	Groundwork of the Metaphysics of Morals
ID	Inaugural Dissertation
Idea	Idea for a Universal History with Cosmopolitan Intent
JL	Jäsche Logic
LP	Lectures on Pedagogy
MF	Metaphysical Foundations of Natural Science
MM	Metaphysics of Morals
NE	A New Elucidation of the First Principles of Metaphysical Cognition
NM	Negative Magnitudes
OP	Opus Postumum
OPA	The Only Possible Argument in Support of a Demonstration of the Existence of God
PM	Physical Monadology
PP	Perpetual Peace

Prog	What Real Progress Has Metaphysics Made in Germany since the Time of Leibniz and Wolff?
Prol	Prolegomena to Any Future Metaphysics
Refl.	Reflexionen
RR	Religion within the Bounds of Mere Reason
RH	Rezensionen zu Johann Gottfried Herders Ideen zur Philosophie der Geschichte der Menschheit
T&P	On the Common Saying: That May Be Correct in Theory, but no Use in Practice
TP	On the Use of Teleological Principles in Philosophy

Acknowledgments

This volume originates from the three-year research activities of an international network grant entitled "Kant and the Laws of Nature: Lessons from the Physical and the Life Sciences of the Eighteenth Century" (IN-081) generously funded by the Leverhulme Trust (2012–2015). Both editors are very grateful to the Leverhulme Trust for making possible the organization of annual workshops at which most of these essays were presented. All of the chapters were written especially for this volume. We are particularly grateful to our network partners on this project (Michael Friedman, Frank James, Peter McLaughlin, Eric Watkins, and Catherine Wilson) for their three-year long commitment to the project and the very many invaluable discussions on the topic of Kant and the laws of nature. Special thanks are due to Jamie Collin, for his sterling work as Network Facilitator for the project and help with proofreading the volume. We are also very grateful to Anna Ortin Nadal for preparing the Index for the volume. Jacqueline Karl at the Berlin-Brandenburgische Akademie der Wissenschaften helpfully provided the cover image of Kant's handwriting. Michela Massimi – as PI for the grant – is very grateful to the Leverhulme Trust for supporting her research and wishes to express her gratitude to the Eidyn Research Centre and the School of Philosophy, Psychology, and Language Sciences at the University of Edinburgh for hosting the project and supporting the various activities related to it. Angela Breitenbach thanks the Leverhulme Trust for supporting her work with a research fellowship, and the Faculty of Philosophy and King's College, University of Cambridge, for hosting the second annual workshop. Both editors are grateful to Hilary Gaskin at Cambridge University Press for enthusiastically supporting this book since its first stages and for seeing it through.

Introduction

Over the years laws of nature have received much attention in philosophy of science, and Kant's philosophy of natural science has become a flourishing area of research. This collection is the first to map the ground of Kant's mature view on the laws of nature. We ask what a law of nature is for Kant; how, on his account, we come to know laws of nature; what necessity laws exhibit; and how laws in the physical sciences might differ from those in the biological sciences. The thirteen chapters collected here shed light on different facets of Kant's account and highlight the centrality of the topic for Kant. As the chapters show, Kant's conception of the laws of nature is continuous with key themes in his metaphysics and epistemology and a core component of his philosophical system.

Kant scholarship has traditionally stressed the differences between Kant's view on laws in physics and biology. The former, expounded most famously in the *Metaphysical Foundations of Natural Science*, find their underpinning in constitutive a priori principles and figure in the mechanical explanations typical of the physical sciences. The latter, famously discussed in the *Critique of Judgment*, find their expression in regulative principles, which guide the teleological considerations afforded by the life sciences. The emphasis traditionally placed on this constitutive–regulative dichotomy – while instructive – has had the effect of obscuring a more profound truth about Kant's view on the laws of nature. For Kant's reflections on this topic are not simply downstream to philosophical discussions about the constitutive principles of the faculty of understanding and the regulative principles of the faculties of reason and reflective judgment. Instead, Kant's reflections on the laws of nature originate from and are informed by his life-long engagement with the sciences of his time. Kant's mature view on the constitutive principles of the understanding and the regulative principles of reason is thus rather the final outcome of his life-long, systematic inquiry into the lawfulness of nature. Moreover, the underlying themes, conceptual commonalities, and metaphysical ideas clearly reveal the continuity Kant saw between the lawfulness of phenomena in the physical and in the life sciences.

Kant was acutely aware that we live in a world that, from the mineral to the animal kingdom, follows regular patterns and manifests lawlike behavior. What then is, for Kant, a law of nature? And in what way do laws govern nature? The first three chapters in Part I set the record straight on these two central questions. In Chapter 1, Eric Watkins charts the territory of the various kinds of law Kant advocates throughout his writings. They include empirical laws of nature; the logical laws of homogeneity, specification, and continuity; and four a priori laws of cosmology, to mention only a few salient examples. This variety notwithstanding, Watkins argues that Kant held a coherent and unified view of what a law of nature is. On Watkins's reading, to be a law means to be necessary, and to be so in virtue of a spontaneous legislative act. On the one hand, Watkins's interpretation stresses the continuity between Kant's mature view in the *Critique of Pure Reason*, where the faculty of understanding is said to "prescribe" laws to nature, and Kant's conception of the moral law in his practical philosophy. On the other hand, Watkins argues that the difference among kinds of laws can be explained by the different cognitive faculties that legislate them, different acts, and ultimately different kinds of necessity.

In Chapter 2, Karl Ameriks probes Watkins's interpretation further, with an eye to underlining the continuity of Kant's theoretical and practical philosophy. On Ameriks's reading of Kant, the necessity of the laws originates from the way in which the antecedent of a lawlike statement acts as a determining ground for the consequent. The determining ground can here be understood as either causal (in natural science) or normative (in morality). By laying out a sophisticated taxonomy with seven main distinctions concerning necessity (and universality) in Kant, Ameriks reminds us of the absolute centrality that the topic of lawfulness plays in Kant, and of its pivotal role to modality, mathematics, and morality.

How we come to know particular causal laws on Kant's account is Paul Guyer's topic in Chapter 3. Hume had brought attention to the limits of what can be known by induction. But Kant's worry about the incompleteness of our knowledge of the laws of nature is not motivated by the same Humean skeptical doubts, according to Guyer. For Kant, we come to know particular causal laws through the workings of the faculty of reflective judgment, in its attempt to fulfill the requirements of systematicity laid out by reason. Our knowledge of particular laws thus depends on our coming to know nature as a *system of laws*, with lower level laws being subsumed under higher level (yet still empirical) laws. On Guyer's reading, it is the systematic unity of the classificatory and explanatory concepts at play in particular causal laws, which ultimately explains why the law-governed behavior of any object is part of a wider law-governed behavior of an entire class of properties (patterns of motion due to gravity, for example). Systematicity plays then an important role for Kant in making us encounter nature as lawful.

Part II of the volume zooms in on the idea of the systematicity of nature and its central role in Kant's account of laws. In Chapter 4, Hannah Ginsborg focuses on the question of why, on Kant's account, we must presuppose the systematicity of nature. She maintains that the principle of systematicity is required not only for pursuing scientific inquiry, but also, and more fundamentally, for arriving at empirical concepts and laws in the first place. Ginsborg argues for this claim by showing that the principle of systematicity is a consequence of the more basic presupposition that nature is purposive for our cognitive faculties. She suggests that we understand this presupposition of nature's purposiveness not as a factual claim about nature, but rather as a claim about the normative fit between nature and our judging of it. On her account, the principle of purposiveness asks us to regard the natural phenomena we seek to understand as being such that our cognitive activity is appropriate to them. The presupposition of the purposiveness of nature is, on this reading, a condition of the exercise of judgment. Furthermore, it commits us to the presupposition that nature is systematic.

In Chapter 5, Rachel Zuckert examines the function Kant attributes to the ideas of reason – the ideas of the soul, the world as a whole, and God – in guiding the search for a systematically unified science. Why, she asks, should a priori conceptions that have no application to empirical phenomena play any role in empirical investigation? Her answer is that, on Kant's account, the ideas function as optimistic placeholders for what there is to be found out in ongoing empirical investigation. On Zuckert's reading, Kant can maintain that the ideas guide empirical inquiry in the search for a systematic science precisely because they do not represent fully determinate, cognizable objects. As nearly empty presentations of objects that lie beyond experience, the ideas encourage us to search for a systematic conception of nature without, however, predetermining empirical science. In so doing, moreover, they indicate the limits of empirical investigation that will never fully satisfy the demands of reason.

In Chapter 6, Thomas Teufel returns to the question of how the principle of the purposiveness of nature relates to the necessity of empirical laws. Teufel makes the case for distinguishing between two different roles the principle plays: transcendental and epistemic. According to Teufel, it is the transcendental function of the principle that wards off cognitive chaos. He suggests that, without presupposing the purposiveness of nature, the threat of a fundamental incompatibility between sensible particularity and conceptual universality would undermine the possibility of human cognition. Furthermore, Teufel argues that the principle entitles us also to consider well-established regularities in nature as necessary and, hence, as laws. The principle of the purposiveness of nature thus serves as the rationale behind the "Maxim of the Lawfulness of Empirical Laws." Teufel construes this maxim as instrumental but not transcendental: it warrants attribution of a material form of necessity to empirical laws of nature.

Nomic necessity and the metaphysics of nature is the overarching theme of the chapters in Part III. If the source of the lawfulness of nature has to be found in the spontaneous legislative acts of our faculties (be it the faculty of understanding, of reason, or of reflective judgment), it might be tempting to read Kant as noncommittal about the metaphysics of laws. Yet, building on recent scholarship, in Chapter 7 James Messina argues that Kant defends a "bottom-up" Necessitation Account of laws of nature, at a distance from traditional readings of both constitutive a priori principles and regulative principles of systematicity. The bonus of the Necessitation Account is that the necessity and universality of the laws can be regarded as supervening on the natures of things (instead of being injected by our cognitive faculties). This reading chimes with contemporary debates in philosophy of science about laws of nature and their nomic necessity. But it also invites us to reconsider Kant's (by and large) underrated metaphysics of nature. By drawing on various primary sources Messina makes the case for a re-evaluation of Kant's metaphysics of nature as central to Kant's Necessitation Account of laws.

In Chapter 8, Michela Massimi sets out to address what she calls the Kantian problem of inference (echoing Bas van Fraassen's objection against David Armstrong's Necessitarian Account of Laws). In its Kantian version, the problem of inference is the problem of explaining how the necessity of effects in nature can be inferred from the Kantian premise that the understanding "prescribes" laws to nature. For it would seem that the understanding could prescribe laws to nature only in a formal sense (qua *natura formaliter spectata*), and not in any genuine real sense. If this (broadly transcendental idealist or projectivist) reading were correct, laws of nature could not possibly be necessary in the genuine sense of prescribing the way nature ought to be. For laws would not have any purchase on nature itself: they would not make any effect in nature necessary. Against this reading, Massimi illustrates three kinds of necessity at play in Kant's account of laws. She explains the nomic necessity of empirical causal laws in terms of a metaphysically more robust picture emerging from Kant's lectures on metaphysics. By showing how nomic necessity ultimately rests on real grounds, and the modal claims associated with it, Massimi shows how Kant is equipped to provide a metaphysically more satisfactory answer to the problem of inference than originally suspected.

The metaphysics of force laws (i.e., the laws of attractive and repulsive forces) is the topic of Daniel Warren's Chapter 9. What grounds these two fundamental laws of nature, according to Kant? Is there any room for experience, given the a priori justification Kant gives to these force laws? Despite their metaphysical foundations, Kant clearly saw the need to supplement metaphysics with a mathematical treatment. Yet metaphysics takes once again a central role as soon as the details of the mathematical treatment turn out to impinge on the very concept of intensive magnitude. What are, for Kant, intensive magnitudes? And to what extent are they amenable to be represented mathematically? For example, are they subject to the mathematics of addition or subtraction (given that they are not extensive magnitudes)? These questions become all the more pressing for the laws of photometry and dynamics. For, in both cases, Kant seems to follow aprioristic lines of reasoning to explain the diffusion of the effects. Yet the very existence of the two fundamental forces cannot itself be known a priori.

Warren's analysis of the interplay between mathematics and metaphysics for the force laws paves the way to Part IV of the volume, dedicated to Kant's conception of laws in physics. Here, two chapters by Michael Friedman and Marius Stan illustrate in detail Kant's mathematization of nature and the resulting necessity of the laws of physics. In Chapter 10 Friedman builds on his previous discussion of causal necessity in Kant by bringing to light the legacy of Newton's methodology of "deduction from phenomena" in Kant and post-Kantian thinkers. On Friedman's reading, Kant endorses Newton's mathematical treatment of gravity as an impressed force and deduces the mathematical properties of such force (i.e., the inverse-square law) from empirical phenomena, such as Kepler's "rules." Kant's conception of force and causal necessity plays an important role in post-Newtonian physical science, in particular in the late-nineteenth-century history of electromagnetism, which eventually led to Einstein's relativity theory.

The legacy of Kant's metaphysical foundations for classical mechanics is Marius Stan's topic in Chapter 11. Despite widespread reports about the untimely death of Kant's metaphysical project in modern science, Stan argues that Kant's "metaphysics of corporeal nature" is still alive and well in modern classical mechanics (if understood primarily as Newton–Euler dynamics). Yet metaphysical problems arise as soon as one considers how modern classical mechanics is amenable to metaphysical treatment in terms of either discrete mass-points or a deformable continuum. Stan deals with both metaphysical difficulties and shows the extent to which Kant's metaphysical foundations can still be embedded in modern classical mechanics.

Finally, in Part V, the discussion turns to Kant's view on laws in the life sciences. Does Kant make room for biological laws? And, if so, what would such laws look like? Angela Breitenbach addresses these questions in Chapter 12 by asking how organic phenomena, famously construed by Kant according to regulative teleological principles, fit with his conception of the lawful unity of nature. Against the widely held view that organisms fall out of the lawful unity of nature on Kant's account, Breitenbach argues that Kant's teleological notion of the organism is compatible with a naturalistic conception of biological entities and, in particular, leaves room for the discovery of genuine biological laws. On her reading, Kant's teleological conception presents a means for identifying parts of nature as organic, thereby guiding the search for biological laws. Relating the Kantian discussion to proposals in the philosophy of biology today, Breitenbach suggests that Kant's approach to the life sciences reveals an idea of the systematic unity of nature whose conceptualization requires a multiplicity of mutually irreducible notions and types of explanation.

In the last chapter, Catherine Wilson traces the historical development of Kant's thinking on the science of living nature from his earliest writings on natural history to his mature theory in the *Critique of Judgment* and his late work in the *Opus Postumum*. Wilson shows how Kant's interest in organic nature leads him to address a number of separate though related problems. They include the ultimate origin of life, animal generation, and the extinction and transformation of species, and are unified by teleological concerns for the purpose of living nature. Against the widely accepted reading that Kant distinguishes the constitutive principles of the mechanical sciences from a regulative teleological approach in the life sciences, Wilson makes the provocative proposal that Kant maintains a constitutive belief in active, organizing forces throughout his career. While in his early writings he regarded such forces as unproblematic, Wilson claims that Kant held on to the reality and efficacy of living forces even when, in his mature theory, he argued that living forces have to be construed according to an analogy with goal-driven, intelligent activity.

As the thirteen chapters in this volume show, Kant had a life-long commitment to the topic of the laws of nature. He came to develop a detailed account of the laws of nature in the physical and life sciences. This account reveals important interconnections between different parts of his vast opus – with similar themes running from the pre-Critical writings to the *Opus postumum*. More important, it reveals a continuity between Kant's theoretical and practical philosophy, and it demonstrates the central role that his metaphysics of nature played for his theory of knowledge. Last but not least, Kant's view on the laws of nature continues to offer a multifaceted and yet coherent image of nature, where the quest for unification does not translate into short-sighted reductionism. This image deeply influenced the course of scientific history after Kant. And it continues to speak to us today and to our ongoing concerns about the systematicity, universality, and necessity of the laws we observe in the physical and biological realms.

PART I

The Lawfulness of Nature

CHAPTER I

Kant on the Unity and Diversity of Laws

Eric Watkins

1.1 Introduction

It is not especially striking, at least at first glance, that Kant makes use of the notion of law in both his practical and his theoretical philosophy. For it is commonplace not only for us, but also for philosophers of the modern period to refer to both the *moral law* and *laws of nature*. At the same time, once one notices just how diverse the kinds of law are that Kant invokes and the very different uses to which he puts them, one could easily be led to suspect that he has no single conception of law that would be coherent and that he is instead playing fast and loose with the notion of law. In this chapter, after surveying a wide range of different kinds of laws that Kant employs, I argue that a univocal notion of law does in fact underlie the different kinds of law that he envisions and that their substantive differences can be explained by the differences in the kinds of necessity, faculty, and act that are involved in laws as well as by the nature of reason, in spite of his clear commitment to the unity of reason.

In the first section, I describe some of the different kinds of laws that Kant accepts, before summarizing, in a second section, the two main elements of a univocal concept of law that underlies these different kinds of law. In the third, fourth, and fifth sections, I then show how the differences between the different kinds of laws can be explained, albeit only in part, by the different kinds of necessities, cognitive faculties, and acts that are involved in these laws. In the sixth section, I consider Kant's understanding of reason's needs, interests, and essential ends so as to be able to provide a fuller account of the differences that obtain between the different kinds of laws. This last consideration has the additional merit of indicating how the different kinds of laws that he employs fit into the aims of his Critical project as a whole (which involves a critique of the entire faculty of reason, both theoretical and practical).

1.2 The Variety of Kinds of Law in Kant's Philosophy

The notion of law (*Gesetz*) plays several highly visible systematic roles within Kant's Critical philosophy. Perhaps the most famous instance of law is found in his practical philosophy, where it finds expression not only in political and legal contexts, where it is most naturally taken to be at home (in the guise of juridical laws or laws of right), but also in the idea of the moral law (*das moralische Gesetz*). While it was common at the time to explain morality, and moral obligation along with it, in terms of a moral law, not everyone sought to do so. Hume, for example, never speaks of a moral law in either his *Treatise of Human Nature* or his *Enquiry Concerning the Principles of Morals*.[1] Further, though Kant devotes much attention to explaining why the moral law must take the form of a Categorical Imperative for finite rational agents such as ourselves, it is also significant that he views the moral law both as the most fundamental principle of morality and as lying at the foundation of what he comes to call the metaphysics of morals.

The notion of law also plays a foundational role in Kant's theoretical philosophy, where it finds expression in the idea of a law of nature (*Naturgesetz*). Again, while it was common by the time Kant wrote the first *Critique* to conceive of what happened in the world as being determined by a small number of laws of nature that are highly general and quantitatively precise, this conception was certainly not prominent or, arguably, even fully present in the ancient and medieval worlds. Indeed, conceiving of science in terms of fundamental laws of nature rather than, say, substantial forms became established only through the path-breaking innovations of leading proponents of the Scientific Revolution such as Descartes and Newton. Further, Kant devoted significant attention to establishing the necessity of a priori laws of nature, arguing in the Second Analogy of Experience, for example, that these laws of nature are necessary for the possibility of experience as well as for the possibility of the objects of experience. The analogy with the moral law extends further insofar as Kant places a priori laws of nature at the foundation of what he calls a metaphysics of nature.[2] As a result of these considerations, it is clear that the notion of an a priori law, whether of morality or of nature, occupies a

[1] Hume does repeatedly refer to laws of justice and civil law but does not seem to conceive of morality in terms of a moral law.

[2] Kant explicitly claims that lawfulness is required for the very possibility of nature in the *Prolegomena* (*Prol* 4:319–320).

foundational place in Kant's overall philosophical system in virtue of being the fundamental principles of the two branches of cognition that constitute his positive metaphysical system.

It would, however, be a mistake to think that this exhausts the conceptual or philosophical space that the notion of law occupies for Kant. He also accepts a wide range of other laws that differ significantly from both the moral law and a priori laws of nature. For example, Kant is also committed to several different kinds of *empirical* laws of nature. In the *Metaphysical Foundations of Natural Science* he argues for three mechanical laws that are closely related to, even if interestingly different from, Newton's three laws of motion.[3] He also accepts the law of universal gravitation and both psychological and chemical laws (*MF* 4:468), which are all empirical in different ways.[4] In addition, in the second half of the *Critique of the Power of Judgment* Kant explores the special status of teleological laws, adopting a complex, but ultimately positive attitude toward them as well (*CJ* 5:409).[5] In fact, in the Introduction to the third *Critique*, Kant develops an account of how one can incorporate a wealth of different empirical laws into a single system in such a way that the necessity of more general laws can flow down into more specific laws. But all of these different empirical laws are distinct from the a priori laws of nature that are at the foundation of the metaphysics of nature Kant envisions not only because they have an empirical content, but also because the way in which they can be justified (by appeal to *actual* experience) must be quite different from the justifications provided for the a priori laws of nature (which appeal only to merely *possible* experience).

However, the laws that enter into Kant's philosophy are not exhausted by empirical laws of nature either. For he introduces the logical "laws" (A653/B681) of "homogeneity, specification, and continuity of forms" (A658/B686) into his theoretical philosophy. These laws state that the understanding must seek (1) more *general* concepts that would represent the sameness of kind for different things, (2) more *specific* concepts that

[3] Cf. also *CJ* 5:387. For fuller discussion, see Watkins 1997 and 1998a. Note that Kant thinks that the Laws of Mechanics have a special status because they are both a priori (insofar as they make experience possible) and empirical (insofar as they presuppose an empirical concept).

[4] I am classifying the subjective laws of logic, which determine how we commonly think (rather than how we ought to think), as psychological laws. In the Blomberg Logic, Kant explicitly remarks: "That science which examines the subjective laws of the rules of our uses of understanding and reason is called *psychologia*" (20:25). Otherwise, Kant grows increasingly skeptical of spiritual laws (*geistige Gesetze*, see *RH* 8:51), which he seemed to be tempted by at moments during his pre-Critical period.

[5] For further discussion, see Watkins 2009.

would represent variety among things falling under the same genus, and (3) a *continuity* of concepts that lie between the more general and the more specific. These three laws are clearly not empirical, since they are not borrowed from experience (B685), but they are also not laws that are constitutive of the possibility of nature itself. Instead, Kant calls them logical laws, though they obviously differ from more standard laws of logic, such as the principle of contradiction, and he suggests that they serve as regulative principles, which have a distinctive status within his theoretical philosophy.

Further, in both the *Critique of Pure Reason* and various transcripts from his metaphysics lectures, Kant endorses four a priori laws of cosmology, which state that there can be (1) no leap, (2) no gap, (3) no fate, and (4) no chance in the world (A228–229/B280–281). These laws, which Kant does not explicitly argue for in the first *Critique*, even if he does make some brief remarks in support of them (A229/B282), state that there can be no empty space or time in the world and that the changes that occur in the world cannot be ruled by fate, cannot be subject to blind necessity, and must be continuous (i.e., cannot admit of discontinuous jumps that would skip intermediate states). Interestingly, although these laws are not empirical, since they are not based on empirical generalizations, they are also not obviously trivial consequences of the Analogies of Experience. Instead, Kant suggests that they are required for the unity of experience (or at least for the unity of the understanding's concepts) (A230/B282).

In sum, what we see when we look into Kant's usage of the term "law" is that he employs several different kinds of law that play a number of extremely diverse roles within his philosophy as a whole. In his practical philosophy, morality takes the form of a moral law that lies at the foundation of his metaphysics of morals, and juridical laws help to constitute the basis of his political philosophy, where both place us under an obligation to act in specific ways. In his theoretical philosophy, he invokes several different kinds of laws that take on different tasks. The idea of a priori laws of nature lies at the foundation of his metaphysics of nature by providing a structure for our experience of spatiotemporal objects, while various kinds of empirical laws – such as the laws of mechanics, chemistry, psychology, and teleology – fill out that structure with the requisite empirical detail. In addition, Kant accepts several nonempirical laws that serve as regulative principles, namely those of homogeneity, specificity, and continuity, and the four a priori cosmological laws of no leap, no gap, no fate, and no chance. In fact, Kant acknowledges even more kinds of laws, since he accepts "objective laws of the understanding and reason," that is, more traditional laws of logic (20:26), as well as what he calls laws of sensibility (24:46), though these will not be my focus here.[6]

1.3 What Is, for Kant, a Law?

Note the incredible diversity here. There is diversity in the kinds of laws to which Kant is committed – laws can be practical or theoretical, a priori or empirical, constitutive or regulative – diversity in the different roles that they play – they can obligate us or govern the world, or serve as normative principles for our thought about it – and finally diversity in the different kinds of justifications Kant provides – transcendental and empirical.[7] In light of this diversity, one might wonder whether it is possible to identify a univocal notion of law that underlies them all or rather whether Kant's use of the term "law" is an incoherent muddle of conflicting notions.

Though a full case would require further argument (Watkins 2014a), I suggest that Kant operates with a univocal concept of law that contains two crucial elements. First, the concept of law requires a necessary rule, where it is the element of necessity that constitutes its distinctive component. It is clear, for example, that both the moral law and the a priori laws of nature are necessary, since the validity of the moral law is in no way contingent (even if it is contingent whether we act in accordance with it) and a priori laws of nature carry with them both strict universality and necessity, given that they are a priori and Kant understands universality and necessity as criteria of a priority. Now empirical laws might seem to be contingent in light of their empirical content, but insofar as they are laws, they too must have an element of necessity, according to Kant. For even if it were contingent whether this or that empirical feature of an object causes a certain effect, it is necessary that there be *some* feature that causes that effect. However, Kant's view is stronger than this, for in the third *Critique* he attempts to show how a system of laws can reveal the necessity of even empirical laws.[8] For example, I take it that for Kant it is necessary that human beings are mortal and that this necessity follows from the necessity of all animals being mortal and of all humans being animals. Thus, every law, regardless of whether it is empirical or a priori, contains an element of necessity.

6 Very rarely, Kant mentions mathematical laws (e.g., at Vienna Logic, 24:802). On logical laws, see Tolley 2006.

7 For further discussion of the nature of the kind of justification Kant provides in the *Metaphysical Foundations of Natural Science*, see Watkins 1998b.

8 See, e.g., Guyer 2005.

Second, because he accepts the basic insight behind the motto "No law without a lawmaker," Kant requires that a law be established by a spontaneous act. More specifically, as is especially clear in the case of political (or positive) laws, a law can be valid only if a proper authority has prescribed it to a particular domain through an appropriate act. In the case of the moral law, for example, Kant holds that reason spontaneously "gives" the moral law in such a way that it applies to and is binding on all rational beings. In the case of a priori laws of nature, Kant analogously holds that the understanding "prescribes" laws to nature, specifically, to appearances such that one can say that the laws of nature govern appearances (or, alternately, that appearances are subject ["*unterworfen*"] to law, i.e., A228/B281, cf. A552/B580). Despite important differences, in both cases, a spontaneous faculty engages in an activity that makes a law necessary, or binding, for some domain that is marked out by a distinguishing feature, whether it be rational beings (the moral law), possible experience (a priori laws), or some subset of appearances (mechanical or chemical laws).

In fact, these two elements of law go together for Kant. For whatever differences there may be between the different kinds of laws that Kant envisions, every law must have an element of necessity, but the necessity of the law must in some way be accounted for, and it is clear that only the spontaneous act of a properly authorized faculty could account for that necessity. In this way, Kant can be committed to a univocal notion of law that includes both an element of necessity and a spontaneous act that gives rise to that necessity.

1.4 Different Kinds of Necessity

This description of the univocal meaning of the term "law" not only makes it possible to see what unifies the different kinds of laws mentioned above, but also suggests three ways of specifying features that distinguish them. For many of their differences arise from (1) the different kinds of *necessity* involved, (2) the different kinds of *faculties* that are responsible for the law and its necessity in each case, and (3) the different kinds of spontaneous *acts* that the faculties in question engage in so as to be responsible for the law. I consider these three differences in turn (in this section and in Sections 1.5 and 1.6).

The notion of necessity involved in all the different kinds of *laws of nature* is that of *determination*. One natural way to illustrate the notion of determination within Kant's system would be with a case of *causal* determination in which, say, two billiard balls collide. When one perfectly elastic body communicates its motion to another perfectly elastic body in impact according to the Laws of Mechanics, the one body *causally determines* an acceleration in the other body just as the second body causally determines a corresponding deceleration in the first. It does so by acting according to its nature (as an extended and impenetrable object endowed with a certain mass), by means of the exercise of its distinctive causal powers (e.g., its repulsive force), and in such a way that its action is necessarily equal and opposite to the reaction of the other body according to the Third Law of Mechanics.[9] In this example, the body that serves as a cause determines, or makes necessary, the effect, but it does so according to the law of the equality of action and reaction, where it is this law that expresses the necessity of the actions that cause the precise changes of motion in the two bodies in such a way that the acceleration and deceleration are equal and necessarily so.

The notion of necessity involved in the case of the moral law is more complicated. For here it is crucial to distinguish, as Kant explicitly does, between the kinds of necessity that obtain in the human and divine cases. In the divine case, the moral law determines what God will do (if God acts at all). Given that the moral law is completely objective for God, God could not have done otherwise, even though God acts freely. As a result, the notion of necessity that is involved in the moral law in the divine case is much the same as that involved in laws of nature. By contrast, for finite (and imperfect) rational beings such as human beings the notion of necessity involved in the moral law is different, since the moral law is expressed in the form of a Categorical Imperative and imperatives do not determine actions. Thus it is possible for us to act contrary to the moral law, even if we ought not to. The necessity that is involved in the moral law for human beings is thus not one of determination, but rather of obligation. Accordingly, when the moral law takes the form of a Categorical Imperative, the necessity of the law is simply obligation, which Kant also calls necessitation:

> *The necessitation of an action by the moral law, is obligation; the necessity of an action from obligation, is duty.* Necessity and necessitation are different: the former is objective necessity. Necessitation is the relation of a law to an imperfect will. In man, the objective necessity of acting in accordance with the moral laws is necessitation. Necessitation is making necessary. (29:611)

[9] For a more detailed description of this kind of case, see Watkins 2005.

In short, the moral law is not necessary for human beings by determining their actions. Instead, it is necessitating by making an action obligatory. As a result, the moral law contains an element of necessity, as all laws must, but it can involve two different kinds of necessity, determination (for the divine case) and necessitation or obligation (for the human case).

Various regulative principles to which Kant is committed seem to have a status similar to that of the Categorical Imperative insofar as both involve normative requirements of reason, but regulative principles are different insofar as they aim at theoretical cognition rather than either action or practical cognition. That is, they do not determine either the world or our thought, but rather represent rational constraints on what we should do, epistemically speaking.

1.5 Different Faculties

Laws can also be differentiated by means of the faculty that is responsible for the act that gives rise to the law and its necessity. According to Kant, reason is the faculty that is responsible for the moral law. This is fitting insofar as Kant characterizes reason as an active and spontaneous faculty of principles, and the moral law, qua law, requires an active spontaneous faculty. However, reason's involvement in the moral law is even more extensive, according to Kant. For reason, as the highest faculty, answers only to itself and thus has the authority to legislate the moral law. Further, reason is responsible both for the content of the moral law and for its necessity. As Kant understands it, the content of the moral law is universal, in the sense of admitting of no exceptions, a feature that is consistent with the fact that reason is the same for all. Reason is also responsible for the binding force of the moral law, because, on Kant's distinctive view, a subject is bound by the moral law only because reason acts autonomously, namely, gives itself the law. That is, Kant's doctrine of autonomy is possible only on the basis of his account of reason, since he understands autonomy not as in any way contingent or as dependent on particular empirical acts, but rather as relying on our basic capacity to reason for ourselves.[10] The moral law is thus grounded in the faculty of reason in several notable respects.

Kant is equally clear that it is the faculty of the understanding that is responsible for the a priori laws of nature.[11] In the *Prolegomena*, he remarks:

[10] See, e.g., Ameriks 2000.

[11] Kant employs both broader and narrower conceptions of reason. Reason in the broad sense encompasses both the understanding and reason in the narrow sense.

"even though it sounds strange at first, it is nonetheless certain, if I say with respect to the universal laws of nature: *the understanding does not draw its* (a priori) *laws from nature, but prescribes them to it*" (*Prol* 4:320). The understanding is in a position to prescribe a priori laws to nature because, like reason, it is an active and spontaneous faculty, though it is unlike reason insofar as its concepts must be applied to objects given in sensible intuition. Further, as the faculty of a priori rules, the understanding can also account for the lawfulness of the laws of nature, that is, for their strict universality and necessity. For the (subjective) rules that the understanding operates in accordance with can give rise to the (objective) rules that are characterized as laws of nature (A126). The understanding is able to prescribe laws to the sensible world in part because it is superior to sensibility (because of its activity and relative independence and the latter's passivity and relative dependence) and in part because, according to Kant's so-called Copernican Revolution, the understanding makes sensible objects conform to it by putting its forms into the object, rather than the other way around.[12]

Empirical laws of nature can be accounted for, in principle, in a way that is analogous to the explanation just given for a priori laws. For empirical laws of nature are a result of the combination of the lawfulness of the a priori laws prescribed by the understanding with an empirical content that has its source in our sensibility. Granted, special issues arise for the complexities pertaining to how sensible content is generated, but even so the understanding has a certain kind of priority over whatever sensible content arises. Now, as noted above, it may be the case that the empirical content of an empirical law can also be shown to be necessary by being entailed by higher-level laws that are discovered through the reflective power of judgment, but that is not essential to their status as laws of nature. Thus empirical laws of nature are based on the faculty of the understanding, just as the a priori laws of nature are, even if they also depend on sensibility.

The laws of homogeneity, specificity, and continuity, by contrast, have a somewhat more complicated origin, though one that is still fully consistent with Kant's general conception of law.[13] On the one hand, considerable evidence suggests that Kant thinks that these laws derive from reason. He

[12] There is, however, an interesting similarity between this and the previous case. In the case of both the moral law and a priori laws of nature, an active faculty gives rise to an a priori form that is prior to an empirical matter, which is subordinate to it, even if the matter is contingent desires in the one case and sensations, or sensible intuition, in the other.

[13] For further discussion, see Watkins 2013.

explicitly states that by means of these principles "reason prepares the field for the understanding" (A657/B685). That is, they all direct the understanding to employ a concept for the sake of a particular explanatory role (of revealing similarities or differences among objects). Also, the laws of homogeneity and specificity seem to have a connection with reason since they aim at totalities that are bound by limits (highest and lowest concepts). Finally, Kant claims that the "continuity of forms is a mere idea for which a corresponding object can by no means be displayed in experience" (A661/B689). That Kant invokes an idea, in his technical sense, clearly suggests that reason is involved.

On the other hand, it is tempting to think that reflective judgment is involved in these laws insofar as they start with a given particular object and then attempt to find different kinds of concepts that cover the particular in question, a process that the faculty of judgment would be responsible for in its reflective use. The differences that are apparent here may be due either to Kant changing his mind or to his supplementing his view from the first to the third *Critique*. Whatever the nature of the changes to his position is, however, the employment of reflective judgment does not preclude the involvement of reason. The faculties of reason and judgment may well complement each other in the case of these laws.

As for the laws of no fate, no chance, no gap, and no leap, two options seem immediately tempting.[14] If we note that these laws are discussed not in the Transcendental Dialectic, which focuses on reason, but rather right after the Postulates of Empirical Thought at the end of the Transcendental Analytic, it would be natural to try to attribute these laws to the understanding. What is unclear about this interpretive option, however, is how these laws are related to the Analogies of Experience (as the most important instances of the Principles of Pure Understanding discussed in the Analytic). They do not seem to follow directly from the Analogies, because their content goes well beyond the claim that, say, every event is caused according to a law. If, by contrast, we take our cue from the fact that Kant discusses these laws in the rational cosmology section of his metaphysics lectures, it would be most natural to assume that they are laws prescribed by reason, though then one wonders why Kant chooses to discuss them in the Transcendental Analytic rather than later in the first *Critique*.

[14] For further discussion, see Watkins 2000.

The best position to take here, I suggest, is to note that while these laws do have their ultimate basis in reason and are thus distinct from the Analogies of Experience, they are explanatory principles that require the Analogies of Experience to do their explanatory work. Thus, if the Second Analogy states that every event is caused according to a law, the principle of no fate goes beyond that by stating what kinds of laws can and, more importantly, cannot be involved, since some kinds of laws would be entirely lacking in explanatory value for us. Specifically, it states that the kind of causal conditioning that is required by the Second Analogy cannot be absolutely necessary, but must rather be only hypothetically necessary (since it presupposes conditions under which it occurs). Regarding, for example, the case of fate, Kant notes: "To want to explain something by destiny is nonsensical, for calling upon destiny just means that I cannot explain something" (29:926). Confirmation for viewing the cosmological laws as arising from reason can be found shortly thereafter: "Destinies conflict with the interest of reason. For if I accept them, then I must renounce the use of reason" (29:926). If this is correct, the cosmological laws are ultimately grounded in reason, which, as active and spontaneous, is able to support these principles being laws.

1.6 Differences in Acts

The different kinds of laws that Kant accepts can also be differentiated by the different acts that are involved. Specifically, in different passages Kant claims that moral and juridical laws arise on the basis of reason engaging in acts of *legislation* (*CPrR* 5:27); laws of nature result from the understanding *prescribing* laws to nature (*Prol* 4:320); while reason *makes demands* at least with the regulative principles of homogeneity, specificity, and continuity (A653/B682).[15] To explain the nature of these different kinds of acts, I briefly describe how Kant understands the conditions under which they occur.

Prima facie, juridical laws are the clearest illustration of the act by means of which a law is enacted. Rousseau, who greatly influenced Kant, held that laws must be enacted by means of an open legislative process that gives rise to the general will through public deliberation and that requires full participation by those who would be subject to the law. Kant's own political philosophy differs from Rousseau's on many points, but he agrees on the importance of a public declaration when it comes to asserting, for example,

[15] See also, e.g., *MF* 4:469, where Kant speaks of the "demands of reason."

that I have a right to an external object as my private property, a declaration that generates an obligation for all others to respect that right, just as I am in turn obligated to respect the property rights established by others' similar declarations.[16] Kant and Rousseau also agree that political laws must be established by citizens endowed with reason, which makes them legislators of the laws to which they are themselves subject. Since the term "law" derives from and is always most at home in political contexts, it is natural to require the public act of a legislator to create an (in this case, legal) obligation.

The case of the moral law is trickier than that of juridical laws, despite the fact that in both cases reason is the relevant faculty and the necessity at issue is one of obligation rather than determination.[17] Part of what makes the moral case more difficult is that some ambiguity surrounds the exact meaning of the notion of legislation when Kant says that reason "legislates" the moral law. What he says, literally, is that reason "gives" the moral law. For example, in the second section of the *Groundwork*, where he is trying to be as clear and explicit as possible, he asserts, in the Gregor translation: "the will is not merely subject to the law but subject to it in such a way that it must be viewed as also giving the law to itself and just because of this first subject to the law" (*G* 4:431).[18] The crucial German phrase here is "*selbstgesetzgebend*," rendered "giving the law to itself." Unfortunately, the meaning of the term "giving" or "*gebend*" is not particularly clear, though when used in the context of law, it is not obviously wrong to translate it as "legislation" (though "lawgiving" is more literal). But note that Kant does not always stick to the term "giving." In one passage he defines autonomy as "a property of the will by which it is a law to itself" (*G* 4:440, cf. also *G* 4:447). Thus, in this case, practical reason (or, what is the same for Kant, the will) does not *give* itself the law but *is* the law, that is, is somehow identical with it. But if the law just *is* reason, how could it require an act?

16 For a recent account of property rights in Kant, see Ripstein 2009.

17 Reath offers an interesting interpretation of the legislation of the moral law: "the parallels are rich enough to warrant Kant's talk of legislating the moral law. The proposal is that we regard the Formula of Universal Law as the 'constitution' of the rational will. It is the fundamental law that sets out the procedure that agents (citizen-legislators) must follow in order to enact substantive principles as law, just as a political constitution sets out the procedure that a sovereign body must follow in order to create law. Substantive moral requirements are the results of the proper application of this procedure, and receive their authority from this fact. When agents guide their deliberations and subsequent actions by the Categorical Imperative, they enact their maxims as law (enact law through their wills)" (2006: 109).

18 "*Der Wille wird also nicht lediglich dem Gesetze unterworfen sondern so unterworfen, daß er auch als selbstgesetzgebend und eben um deswillen allererst dem Gesetze . . . unterworfen angesehen werden muß.*"

Further, in the case of the moral law reason's "legislation" may not involve a public declaration. For one, the moral law obligates regardless of whether anyone ever says so. For another, if the moral law is stated publicly (or commanded by God), what one is obliged to do could not be different after the public statement from what it is before. For both the content and necessity of the moral law do not depend on anyone acting at a particular moment in time and do not require a public act of endorsement, as is the case for juridical laws. Instead, the moral law is always binding, regardless of what one actually says or does at any moment in time. But if this is the case, the notion of act involved in reason's legislation of the moral law cannot be any ordinary empirical act.

Similar issues arise for the case of the understanding and laws of nature. Though Kant states (in both the *Prolegomena* and the first *Critique*) that the understanding prescribes laws to nature, he clearly does not mean to say that particular empirical subjects must prescribe laws to nature before particular empirical laws obtain for the world of appearances.[19] These acts of prescription are not datable historical events, in the absence of which the spatiotemporal world would be a lawless chaos. Instead, the laws of nature obtain, regardless of what particular act one particular person's understanding performs. But without empirical subjects engaged in particular acts at particular moments in time, it is natural to wonder what it means for the understanding to be *prescribing* laws to nature, just as the question arises for reason's legislation of the moral law.[20]

What these considerations suggest is that the acts that are supposed to give rise to the different kinds of laws are a matter not of empirical but rather of transcendental psychology. While one might follow Strawson in rejecting transcendental psychology as hopelessly confused and unnecessary, that would, I think, create serious obstacles to understanding Kant's actual position. For the transcendental aspect of transcendental psychology can be quite useful for understanding the normative dimension of Kant's philosophy. As a result, to understand Kant's position on laws one must attend to the normative dimension of the acts that are instantiated under particular empirical conditions, and note that the normativity derives not from the empirical conditions, but rather from the principles that guide the activities that take place under empirical conditions. Thus, instead of

[19] Such acts can be required for actual cognition.

[20] The regulative laws of homogeneity, specificity, and continuity and the four cosmological laws are easier to accommodate on this particular point, since they involve explanatory principles that are specific to finite rational cognizers.

thinking of the relevant act as an efficient cause of the law, it is better understood as what one might think of as a formal cause or constituent principle of law.

If we pursue this line of thought, we can see that reason and the understanding are spontaneous faculties performing "acts" that are responsible for unifying a plurality of one kind into a unity of another kind. The understanding unifies the sensible manifold given in intuition under concepts in a judgment that can be true or false, while reason, at least in its logical use, unifies a plurality of cognitions into syllogisms. But what is crucial here is that these faculties are guided in their unifying acts by certain "functions" or normative rules. The understanding is guided by the functions that find expression in the table of judgments and the table of categories, while reason is guided, in its purely logical use, by what we would call logically valid inference rules. Thus, the acts are crucial insofar as they are responsible for, or are at least indicative of, a distinctive kind of normativity in our judgments about the world. This is consistent with the idea that laws of nature are said to govern the world and the moral law is a normative standard for rational behavior. Thus the normativity that these faculties involve in their acts is a crucial feature of the laws that Kant is trying to explain. Unless one rests content with brute normative laws, one will need a story about what the ground of normativity is in a world that can seem to be populated entirely by purely natural entities, as well as how and why they are normative principles for beings like us. Kant's account involves acts that will happen under various empirical conditions, but what is crucial about the acts is not so much that they unify something at a particular moment in time as that the unity they result in embodies certain functions that allow them to support normative principles.

Though this account of acts and their constitutive functions captures a distinctive and important part of Kant's views on laws, it still leaves unexplained certain features of laws and of the different kinds of laws that he envisions. For example, though reason is responsible for the legislation of moral and juridical laws as well as for the demands of the principles of homogeneity, specificity, and continuity, reason has only one set of unifying functions on which to rely. But how can one set of functions account for the different kinds of laws that reason is responsible for? It cannot be a difference in faculty, since reason is involved in all of these acts. Nor can we appeal to a different set of functions, given the unity of reason on which Kant insists.[21] The notion of necessity is different in some

[21] See, e.g., *CPrR* 5:89 and *G* 4:392 for explicit statements (and endorsements) of the unity of reason.

of these cases, but only in some, and that is a *result* of the different kinds of acts, not the *source* of their differences. To make progress on this problem, we must consider, very briefly, the nature of reason.

1.7 Reason's Needs, Interests, and Ends and the Project of the *Critique of Pure Reason*

Given Kant's commitment to both the unity of reason and the univocal concept of law, how can he account for the different kinds of acts that give rise to different kinds of laws, with their different kinds of necessity? Part of the answer stems from reason having two distinct objects. As Kant points out in the Doctrine of Method's Architectonic of Pure Reason: "The legislation of human reason (philosophy) has two objects, nature and freedom, and thus contains the natural law [*Naturgesetz*] as well as the moral law, initially in two separate systems but ultimately in a single philosophical system" (A840/B868). What this passage makes clear is that the difference between reason's legislation of juridical laws and the moral law, on the one hand, and its activities in theoretical philosophy, on the other, derives from a difference in its object, since in the former case its object is freedom, while in the latter its object is nature. Since reason has different objects, it is not surprising that its laws with respect to those objects will be different as well.[22] At the same time, the difference in reason's objects cannot account for all of the differences between the different laws, but rather only for the basic distinction between practical and theoretical laws. As a result, one must appeal to something other than reason's object to account for the further differences we find between the different kinds of laws.

The most significant factor, I want to suggest, lies in the fact that Kant takes reason to have needs, interests, and essential ends.[23] Consider first Kant's view on the needs of reason (which contrasts with the needs of inclination). In one of his more ambitious moments Kant claims that reason should be able to achieve "insight into the unity of the whole pure rational faculty (theoretical as well as practical) and to derive everything from one principle" (*CPrR* 5:91). This ambition is based, he thinks, on "the undeniable need of human reason, which finds complete satisfaction only in a complete systematic unity of its cognitions" (ibid.). That is, reason has

[22] This difference is also responsible for Kant's division of metaphysics into a metaphysics of freedom and a metaphysics of nature.

[23] See Kleingeld 1998 for discussion of the conative character of reason.

a "need" to derive all cognitions, whether theoretical or practical, from a single principle and it can satisfy this need only by developing a complete systematic unity of cognition. In short, reason has a need for systematic unity.

Now "interest" is a technical term for Kant, which indicates a need of which a rational being is also aware. That is, if I am rational and aware of a need, I have an interest in satisfying that need. Kant then employs this notion in different contexts. For example, he invokes it when he claims that "all interest of my reason (the speculative as well as the practical) is united" (A800/B828) in the three questions about what I can know, what I should do, and what I may hope. Thus, we not only have a need for a complete systematic unity of cognition, but are also interested in precisely that complete systematic unity of cognition that would allow us to answer these basic questions. At the same time, he uses the notion of an interest in more specific contexts. For example, he notes that the principles of homogeneity and specificity express "two interests that conflict with each other" (A654/B682), since the one pushes toward ever more generality while the other pushes toward ever greater specificity. However, he goes on to resolve this apparent conflict by noting that "reason in fact has only a single unified interest and the conflict between its maxims is only a variation and reciprocal limitation of the methods satisfying this interest" (A666/B694). Though Kant is not explicit about what the "single unified interest" is, I take it that having both increasingly general and ever more specific concepts is necessary to have the richest possible set of concepts, which is in turn necessary to achieve a *complete* systematic unity of cognition.

But the crucial point about these needs and interests is their relation to the essential *ends* of reason. For a need of reason (and therefore every interest as well) presupposes an end of reason. For only if my end is not satisfied do I have a need. In fact, the nature of the need can be determined by the nature of the end and what resources would be needed so that the end can be accomplished.

The fundamental question then is what the essential end of reason is. Kant asserts that "reason demands to cognize the unconditioned, and with it the totality of all conditions, for otherwise, it does not cease to question, just as if nothing had been answered yet" (20:326).[24] That is, reason's essential end is to cognize not only the conditions for whatever conditioned items are given, but also the totality of these conditions, and

[24] See also 20:287.

therefore the unconditioned, since cognition of the unconditioned alone represents a resting place where its needs and interests have been satisfied. In theoretical philosophy reason pursues its essential end by attempting to achieve theoretical cognition of God, the world as a totality, and our soul, since these objects of traditional metaphysics are unconditioned. In practical philosophy, reason pursues its essential end by cognizing either the moral law, which binds categorically and thus unconditionally, or the highest good, which contains virtue, an unconditioned good, and, proportionate to it, happiness, which is good only on the condition that one's actions are virtuous enough to make one worthy of happiness. By conceiving of the essential ends of reason in terms of the unconditioned and then understanding the unconditioned along both theoretical and practical dimensions, Kant is able to assert the possibility of a complete systematic unity of cognition while still maintaining the unity of reason, despite obvious differences between the objects of reason (nature and freedom).

The *Critique of Pure Reason* and the *Critique of Practical Reason* then undertake full-scale investigations of the nature and limits of our cognitive capacities so as to determine whether we are in a position to attain the kind of cognition that could be adequate to reason's essential ends. As it turns out, detailed analysis of our theoretical faculties of sensibility and understanding reveals the limitations that prevent us from having theoretical cognition of the unconditioned objects of traditional metaphysics so that reason's end cannot be attained on purely theoretical grounds. Fortunately, however, a similarly detailed analysis of our practical capacities of desire and will reveals that we *can* have practical cognition both of the moral law, along with the freedom that is a prerequisite of it, and of the possibility of the highest good, along with its presuppositions of God's existence and the immortality of the soul. As a result, the "self-cognition" (Axi) that reason acquires through its investigation of its cognitive capacities establishes that its own ends can ultimately be fully satisfied on practical grounds.

But what emerges out of this result, which is of direct importance for our question about Kant's understanding of laws, is that the different kinds of laws that reason is responsible for, and the different kinds of acts that reason engages in, derive from several factors. Reason's different objects, nature and freedom, constitute one factor. But another lies in the different cognitive tasks that arise when reason attempts to gain complete systematic cognition of something unconditioned with the limited cognitive resources at its disposal. Accordingly, the a priori laws of nature are required for the constitution of the possibility of experience, which is the form that theoretical cognition takes for beings endowed with a priori forms of sensibility and understanding. The empirical laws of nature are required for the specifically empirical cognitions that fill out the a priori structure established by the pure forms of sensibility and the understanding. Principles of reflective judgment are required for the discovery and justification of empirical laws since the complete system of cognition requires that the necessity of the empirical laws be justified and the principles of reflective judgment are required for that. The laws of homogeneity, specification, and continuity are required because the completeness of the systematic unity of cognition could not be achieved if our cognition did not involve the richest possible conceptual resources, and these laws direct the understanding to develop and deploy the fullest possible range of concepts. The laws of rational cosmology – no gap, no leap, no chance, and no fate – are required because they express the explanatory demands that must be met for the systematic unity of cognition to be as rich as possible so that it can approximate completeness. That is, the point of a system of cognitions is not simply to relate all cognitions to each other in some way, but rather to relate them to each other in such a way that both the whole and its various parts become intelligible in virtue of the interrelations of the parts and the idea of the whole. The laws of rational cosmology are thus relevant insofar as they help to ensure that the whole of cognition satisfies these expectations. In this way, each of these particular kinds of laws makes a different kind of contribution to the complete systematic unity of cognition that is reason's ultimate end.

1.8 Conclusion

It is, of course, possible that laws of nature are not really laws at all, or are not laws in the same sense as other kinds of laws. And perhaps morality is not best expressed, or expressed at all, in terms of a moral law. Perhaps, if Hume and his followers are correct, laws contain no necessity and neither govern the world nor generate obligation, but rather are simply contingent descriptions of what happens in the world, or express our attitudes. Instead of attempting to refute such a view, I have provided a fuller description of Kant's account of laws, one that captures many of the different kinds of laws to which he is committed. I have done so in the hopes of showing both that there is a univocal notion of law that they all share, despite their differences, and, especially, that Kant can account for the differences between them by appealing to the resources that he develops in the course of working out his overall philosophical project in the three Critiques.

Again, perhaps Kant's larger view of things is untenable, but even if that should be the case, it is instructive to see what a robust alternative picture might look like, one that fully embraces his initially strange-sounding commitments, instead of shying away from them.

Acknowledgments

I thank Clinton Tolley and the audience at the Leverhulme conference held in Edinburgh in June 2015 for helpful discussion of and comments on an earlier version of this chapter. Special thanks go to Michela Massimi and Angela Breitenbach for organizing the project on *Kant and the Laws of Nature* of which this conference was one central part.

CHAPTER 2

On Universality, Necessity, and Law in General in Kant

Karl Ameriks

2.1 Some General Distinctions

In addition to the close links that Kant frequently draws between the notions of universality, necessity, and law, he draws a fundamental distinction in the meaning of these terms between (1a) "*true or strict*" and (1b) *loose* or "comparative" understandings – as in mere regularities or "assumed" constructions that cover actual situations but not what we call lawful hypothetical ones (B3; cf. A2 on "inner necessity"). Unfortunately, Kant also often fails to make this distinction in an explicit way, and he frequently writes in a shorthand manner that relies for convenience on the fact that an exception to claims of universality can be more easily perceived than an exception to claims of necessity. This tendency has led many readers to get entangled in problems about popular notions of universality and conventional law that are unmoored from a recognition of Kant's primary concern with the strict notion of necessity.

As a consequence, many well-known but ill-founded objections to his philosophy have arisen, objections that can be countered by noting ways in which the full context of Kant's arguments reveals that his own basic use of all these terms should not be understood in terms of a loose sense that would imply the absurd subjectivist consequences that unsympathetic interpreters sometimes attach to his philosophy.[1] This point is best appreciated by tracing the paradigmatic meaning of Kant's notion of law through a wide range of passages, not only in the first *Critique* but also throughout a variety of other texts, including writings in all periods concerning practical issues. These writings clearly show a dominant attachment to an across-the-board strict understanding of universality, necessity, and law, one that fits not only the general formal meaning used in logic but also the substantive meaning found,

[1] See Ameriks 2006: chapter 6.

for example, in the notion of the moral law as a principle that holds for even a perfect divine being. There are further complications, however, such as the fact that, in his Critical theoretical philosophy, Kant uses a substantive in-between notion of law that nondogmatically affirms, for the empirical domain, transcendental principles that are not comparative or subjective but are also not meant to entail necessities in an absolute metaphysical sense – what Kant calls "intelligible" necessity (A564/B592). In view of complications like this, it is useful to construct an *elementary taxonomy* of some of Kant's other general distinctions, simply as a relevant first step toward any systematic evaluation of the more controversial specific claims involving necessity that Kant eventually makes concerning natural science and its methodology, especially with respect to the complicated domain of the organic world.

A second fundamental distinction here, one that should not be conflated with the just noted contrast in the basic *meaning* of modal terms, concerns a difference in *scope*, namely, between (2a) the relatively *strong* assertion of *particular* special rules as, in some sense, *individually necessary* and perhaps even certainly, demonstrably, and nonregressively so, and (2b) the relatively *weak* regressive and transcendental claim that, given the basic premise of the possibility of experience, it is *in general necessary* for us to rely on *some* objective rules. For example, it can be (2b) necessary for modern traffic or the grammar of a particular language that there be some rules or other (so that people don't just drive or talk in any way) but that does not mean or entail that (2a) a particular basic traffic rule (e.g., drive on the right) or natural language grammar rule is in itself necessary. This distinction is highly relevant to but hardly elaborated in Kant's brief introductory comment at B5: "For whence could [even] experience derive its certainty [note that this is a premise, not a conclusion], if *all* [emphasis added] the rules, according to which it proceeds, were *always* [emphasis added] themselves empirical, and therefore contingent."

Note that, whether or not Kant intended it this way, one can identify a twofold use of the notion of universality (and hence also of necessity) in this remark, and this naturally invites the thought of a couple of further options, namely, alternatives of an in-between kind where, in each case, one, but only one, of the universal claims is dropped. Thus, between (2a) and (2b), there arises, from a merely formal perspective, first, the alternative of saying that, even if not all objective rules can be always empirical, still at least *some* of them might be always empirical,[2] as well as the alternative that perhaps all rules (i.e., any individual rule) might be *sometimes* empirical (even if it is also maintained that this possibility cannot be actualized for all rules all at once). Presumably Kant himself, however, would not pursue the latter formal alternative, for he seems to presume that (for any world we can experience) if any rule – for example, the characterization of matter – is actually a priori or necessary at all, then it will be so in all *such* worlds. Without deciding about the impact of this debated complication, which becomes highly relevant in later neo-Kantian philosophies where rules may vary in their a priori status in different systems, it suffices for now to keep in mind that, at the very least, the fundamental distinction between (2a) and (2b) is indisputably of relevance within the Critical philosophy. One need only recall recent controversies about how the Second Analogy is to be read, where some readers[3] take its concern to be the weak claim, "Given experience, necessarily there are known causes," while others take it to involve the strong claim "There are causes known as necessary."

A similar perplexity arises not only with the other Analogies but also in Kant's practical philosophy. In the *Groundwork* Kant begins by taking for granted that our conditioned value assertions ultimately depend on an unconditioned value (*G* 4:393f). He does not go on to explain that in the first *Critique* (A417/B445) his own discussion of reason's notion of the unconditioned highlights the point that, even when our reason requires us to posit something unconditioned, this unconditioned may take the form, not of a particular first ground, but of a sequence of grounds as a whole – a whole that in the context of the *Groundwork* would not be characterized in spatiotemporal terms. Hence its assertion need not run afoul of the antinomies that arise in the *Critique.* Kant's not exploring this option may seem to imply that his greatest concern is not just any unconditional "absolute" but simply the unconditional *causal* responsibility of finite *individuals*, whose free choices function as absolute first grounds of action. In fact, however, it is not unconditioned *individual causal* responsibility alone that matters for Kant, but this responsibility in the face of an absolute value – namely, concrete persons as ends in themselves, and not

[2] See Friedman 2014: 533, "what is most problematic . . . is the suggestion that all experience . . . must nevertheless involve some kind of necessary connection." Friedman goes on (545) to point out how Kant later turns to considerations of systematicity to try to secure the presence of necessary rules throughout experience.

[3] See Michael Friedman's contrast of the "virtually unanimous" stress of interpreters on the "universal principle" every event has a cause, with his own stress on (Kant's commitment to) "particular instantiations of the claim that all events of type A are followed by events of type B" (1992b: 164).

a mere entire series of conditioned values without end – that involves the *general* form of a *law* of unconditional ("categorical") *significance*.[4] Without such a value, our existence, even when it is characterized in libertarian causal terms and is not thought to suffer the fate of a "mere turnspit," would still take the form of what Kant repeatedly ridicules as "mere play."[5]

In view of this primacy given to morality, my overall hypothesis is that Kant may have felt it was not *most* important to defend in detail a strong (but not dogmatic) reading of the scope of necessity in his *theoretical* philosophy as long as he believed that a strict reading is clearly needed, *and* that a strong reading of some unconditioned norm can be relatively easily affirmed in our practical life. The affirmation of this norm is, after all, meant to rely not on complex and controversial theoretical claims but on what is supposedly implicit in all "sound common understanding." Notes from Kant's reading of Rousseau in the 1760s already show that his philosophy is dominated (albeit, for decades prior to the *Groundwork*, only in numerous notes and lectures rather than publications) in general by the strong claim that there is an overriding and necessary moral law, understandable by each of us as applying equally to all of us, rather than the weak claim that necessarily at each moment there must be some principle(s) or other (such as prudence or sympathy) guiding humanity's practical life.[6]

That these issues can have a variety of implications at different levels becomes obvious when one distinguishes, third, the fundamental *fields* of necessary claims that concern Kant. Like other philosophers, he is, of course, concerned with (3a) *formal logic*'s narrow and analytic notion of necessity – which remains independent of the doctrine of transcendental idealism – as well as with (3b) assertions of concrete necessity in *traditional metaphysics*, as in arguments using the principle of sufficient reason (albeit from a Critical epistemological perspective). What is positively distinctive about his Critical philosophy, however, is its stress on (3c) *transcendental philosophy*'s unique notion of experience-constituting necessities, that is, principles expressing universal laws for minds that are, broadly speaking, *epistemically like* our species. Here the notion of "our species," like the philosophical notion of "human," is to be taken in a general sense and not literally as a narrow biological term that refers, at most, to *Homo sapiens* and its immediate earthly relatives. This is because Kant clearly believes that some of what he is arguing for in the *Critique* can also apply to rational beings in other solar systems that have a composition and origin different from ours.

4 See Ameriks 2012: chapter 8. 5 See Ameriks 2012: 23, n. 28. 6 See Ameriks 2012: chapter 1.

In sum, and looking back on all three basic distinctions introduced so far, a preliminary conclusion to draw is that both strict and strong understandings of modality clearly seem most appropriate for Kant's understanding of logical and traditional broadly metaphysical principles, and what remains unsettled is whether (and if so, why) a similar double understanding has to be adopted for all asserted transcendental necessities.

At this point a fourth fundamental distinction needs to be introduced, for Kant's (4a) *human* transcendental necessities need to be seen in contrast not only to what is distinctive about the highest (divine) and lowest (brute) types of mind, but also to (4b) possible *nonhuman* transcendental necessities for minds that are finite, receptive, and rational. These necessities would apply to minds in other possible worlds that are in a position that is *like* ours insofar as that position is also in between the divine and brute situations, and yet is also very *unlike* that of our epistemic species insofar as it lacks the specific kind of a priori rules that govern our distinctive forms of sensory data. For such minds, Kant repeatedly stresses that the specific human rules of the first *Critique* – the spatiotemporally schematized principles of the Transcendental Analytic – are not relevant, although for these minds there still might be analogous rules defining different instances of transcendental necessity, that is, a different sensibly schematized use (its "synthesis speciosa," B151) of the same categories that define our pure understanding.[7] Whatever decision is made then about whether and how Kant is, or should be, committed to what has just been characterized as strict and strong transcendental necessities *for us* must therefore take into account the fact that it does not immediately follow that these specific principles are intended to apply to finite receptive minds *in general*, let alone for whatever is meant by "things in themselves."[8] Here, as in many other contexts, it is important to keep in mind that, even if *in this sense* there is something that

[7] By "nonhuman" I mean "not having spatiotemporal sensibility" since this is what for him philosophically distinguishes us as a species. Kant admits that we cannot say positively what a nonspatiotemporal sensibility would be like, since we are only human, but the thought of the logical possibility of such a sensibility is crucial to his point that although there is no way to think apart from the intellectual forms of the categories, we must allow that there might be, for other beings (angels, other finite spirits), other ways to sense than the a priori forms of intuition we happen to have, and so, contra Leibnizians, these forms are to be regarded as metaphysically contingent and not in principle derivable from the concepts of things in themselves.

[8] Moreover, the conditions of finitude and receptivity need to be distinguished, for there could be epistemically receptive beings that are in some ways nonfinite, as well as epistemically nonreceptive beings that are in some sense still finite, and each of these kinds of minds might involve different transcendental conditions. Cf. Pollok 2014: 510f., who cites *Metaphysik Mrongovius* 29:800 and *ID* §6 as typical expressions of Kant's doctrine that we are receptive beings with respect to "matter," but not with respect to "form," since this cannot be "given."

Kant speaks of as "subjective" or limited, about our specific principles, this does not mean that they are not at all to be understood as objective as well, in what a Strawsonian would call a genuine "weighty" sense.

Fifth, there is also a need to distinguish, even simply in our own case, between what in general Kant calls (5a) "constitutive" and (5b) "regulative" necessities (for more details on this distinction, see below),[9] as well as, sixth, between these kinds of necessity in relation to different operations of our intellectual *faculties*, which operate at various levels of complexity. For example, (6a) the conditions of *mere judgment* as such (which can be characterized further in terms of simple or complex relational forms, as well as "determinative" or "reflective" ones) contrast with (6b), the richer conditions necessary specifically for the process of forming, and then developing, a complete *system* of scientific judgments. A scientific system can in turn be characterized as either relatively weakly or strongly interconnected. In addition, the specific conditions of judgment and of a system need to be contrasted with (6c), the conditions for satisfying various demands specific to *reason* as our highest discursive faculty.[10] With regard to both (6b) and (6c), there remains the obvious objection that relatively weak and/or loose alternatives may well seem in fact sufficient for *some* successful use of our capacity for systematization and even for what most philosophers may call reason.[11] In that case, few, if any, *particular* strong claims of transcendental necessity at these levels, especially in a constitutive sense, would survive as clearly essential (or at least unrevisable in all parts) for us, let alone other kinds of minds – even if the general weak but still broadly transcendental claim is let stand that we (and our unknown transcendental relatives) always need to rely on some kind of rules if we are to make assertions that are to be regarded as warranted (this position can be called late-Wittgensteinian neo-Kantianism.)

Here it would be nice, of course, to be able easily to identify at least a few contexts in which a Kantian can confidently mount a counterattack against the alternative of extreme forms of creeping pragmatism, without retreating to an overly dogmatic theoretical invocation of transcendental

[9] See McLaughlin 2014a.

[10] Terminological complications arise here because at times Kant also uses the term "reason" to signify the faculty of intellect in general, in contrast to sense, rather than merely the highest, and unconditionally demanding, level of that general faculty.

[11] See Chignell (2014: 593) on the difficulty in determining what is "really possible" for Kant, and in saying, without recourse to empirical considerations of systematicity, why some nonactual concepts, such as telepathy (in contrast to magnetism) are supposedly ruled out *necessarily*, i.e., as transcendentally empty for us.

necessities.[12] My immediate goal, however, is not to attempt a detailed apology for the Critical philosophy on this difficult point but simply to draw attention to how multifarious are the questions that still appear unresolved with regard to the many very different aspects of Kant's position. The objective here is not so much to settle one particular issue as to provide a preliminary overview that may be of help in eventually determining some kind of broad but clearly defined demarcation of what on the whole now seems generally understandable and promising in transcendentalism, and what still remains fundamentally unclear and/or unattractive in Kant's overall position on law and necessity.

Given such a demarcation, one could then more efficiently explore the implications of Kant's overall position for what, in English, can be abbreviated, seventh, as the enduring problem of the multiple mystery of our four key M terms: *modality*, *math*, *morality*, and *mind*. This frustrating quartet roughly corresponds to Kant's division of Critical philosophy into the three *law-governed topic* areas of (7a) general logic, with its analytic laws and formal treatment of *modality*; (7b) metaphysics "that can come forth as a science," which provides foundations for natural science in synthetic transcendental principles of spatiotemporal nature – principles that for the Critical Kant must involve a reference to what is *mathematizable*; and (7c) ethics, which lays out the foundations and implications for us of synthetic categorical imperatives grounded in an absolutely universal *moral* law; along with, finally, (7d) the additional mystery of *mind*, understood as corresponding to the higher-level issue of how our subjectivity can make unified sense of all these topics in a way that is at least compatible with the *Critique*'s ultimate version of transcendental idealism.

Given the three basic levels distinguished earlier (as 6a, 6b, and 6c) in the operation of our cognitive faculties – judgment, system, and reason – there already arises, even apart from the special Kantian problem of mind and idealism, a complex matrix of issues. The three faculty levels are relevant as transcendental conditions for dealing with each of the three law-governed topic areas – the modal, the mathematical, and the moral – and hence, given the additional constitutive/regulative distinction, there are at least 18 times (=2 [from 5a and 5b] × 3 [from 6a–c] × 3 [from 7a–c]) that the complex question can be raised of whether – and then why – a strict and/or strong assertion of transcendental necessity is required; and hence, there are at least 72 sets of possible answers. To sum up, we have seen so far seven main distinctions that can be found in Kant on necessity:

[12] See O'Shea 1997 and Kreines 2009.

1a strictly universal *meaning*	1b vs. loose (theoretical and practical)	
2a relatively strong *scope*	2b vs. weak (theoretical and practical)	
3a *fields* of logic	3b vs. traditional metaphysics	3c transcendental conditions
4a *transcendentally* necessary for us	4b vs. other finite (and/or receptive) minds	
5a *constitutively* necessary	5b vs. regulatively	
6a necessary for *faculty* of judgment	6b vs. specifically for system	6c or reason
7a for *topic* of law re modality (logic)	7b re math in nature	7c re morality and 7d re mind

With respect to all these contexts, the ultimate Critical question that arises is, eighth, whether the *ground* of necessity, universality, and law, in Kant's paradigmatic sense, is to be traced to features that (8a) we *antecedently* understand about mind as such, that is, in Kant's unique sense of transcendental subjectivity, in a way that at first may or may not require transcendental idealism, or (8b) whether, conversely, Kant in effect holds that the best that we can say *philosophically* about mind is to be gleaned from the implications of an antecedent grasp of the lawfulness that we become aware of in our use of the principles of (logical) modality, mathematics (and its concrete realm), and morality. If the latter is the case – which is my hypothesis – then this makes it all the more important to determine exactly *how* the three topic areas of lawfulness are related to each of the three basic levels of faculty complexity. In approaching this question, it would be good to be able to determine, first, whether there is some form of univocal and acceptable common Kantian meaning of law, and perhaps also some kind of hierarchy in uses of the term, so that there are good reasons for regarding some contexts of lawfulness as more fundamental for us than others, and for holding on to relatively strong as well as strict assertions of necessity.

2.2 Incorporating Watkins's Insight: A General Kantian Notion of Law

One very helpful response to this issue is offered in Eric Watkins's essay, "What Is, for Kant, a Law of Nature?," an essay that covers more than its title directly expresses, for it argues in detail that there is a common and strict notion of law in Kant's theoretical and practical philosophy. The notion is characterized by the three features of being (Wa) "a necessary rule" that is (Wb) "prescribed by a proper authority" to (Wc) "either determine" a state of affairs or "give rise to an obligation" (Watkins 2014a: 474). This characterization is difficult to improve on, but some minor adjustments and clarifications are worth considering.

One way to simplify the third condition is to focus on Kant's key notion of a "*Bestimmungsgrund*," which plays on the multiple sense of the German term "*bestimmen*" – corresponding to a double sense of the English term "determine" – and can signify a determination in either a mere theoretical characterization of a natural state or the rationalization of an action in a practical context.[13] The latter sense is crucial for Kant's speaking of our actions as "determined" by reason, even though – or, better, precisely because – he holds that reason does not function as a *deterministic force*. Hence, "*Bestimmungsgrund*" is best translated not by the psychological and quasi-mechanistic word "motive" but by a term with a normative implication, such as "ground" meant in the sense of the objective reason with which an agent intends and justifies an action, in contrast to a mere incentive ("*Triebfeder*"), which subjectively accompanies it as a feeling in the agent's mind.

We can now slightly simplify the characterization of feature (Wc) by saying that *both* theoretical and practical contexts involve a kind of *determination* and that in this context the reference specifically to obligation, which Watkins builds into his general account of law, is not *fundamental*. This is because the situation of obligation is something that arises only as the result of a combination of the more fundamental practical notion of moral law, meant in the specific *normative* sense of a determination that involves an absolutely necessary and universal value, with the ultimately contingent situation of free receptive beings such as us, whose needs (qua sensible beings) tempt them to ignore the law. Although obligation for Kant is thus not an intrinsic feature of lawfulness, it is still a "fact of reason" that, given the situation that human beings are in, they are obliged to follow the law, as individuals subject to it and capable of dutifully responding to its commanding demand, even though in an everyday *factual sense* they do not always do so, and they are never (metaphysically) necessarily determined to do so.

One key passage that Watkins relies on here, from notes in the Kaehler lectures, provides a pithy general characterization, which, given the dual meaning of the key term "action" (*Handlung*), is clearly meant to apply to

[13] For further ambiguities, see Ameriks 2016.

theoretical as well as practical contexts of determination: "a formula that expresses the necessity of an action is a law."[14] The obvious twofold general relevance of Kant's notion of a formula depends on an important ninth feature of necessity, namely, its applicability in both (9a) *causal* and (9b) *normative* contexts. (See *CPrR* 5:8 note, where Kant stresses the significance of being able to find a "formula" for morality). This is a very significant contrast for Kant because, in the writings of the 1780s and even before then in nonpublished contexts, his most pressing concern is to find philosophical room for the idea of morally significant action – either dutiful or evil – that involves a normatively necessary determination – a "*Bestimmung*" that can also mean "vocation" or "calling" – that is precisely not absolutely causally necessary, even though it is in accord with laws of nature and may be in principle predictable.[15] Furthermore, the Kaehler passage, with its stress on the term "formula," can also be understood as implicitly playing on the key Kantian notion of *universalizable* determination, since a formula typically is a general expression most efficiently detailing, and in that sense determining, derivations concerning what *always* is true, or happens, or is supposed to happen, when specified conditions are met.[16] Watkins's three-feature account can thus be slightly expanded into a *general* characterization of the notion of law through a direct use of a common concept of determination (in contrast to his phrase "rather than determining"; Watkins 2014a: 486). The manifold senses of this concept make it especially appropriate for revealing the very close connections between all three general notions of law, necessity, and universality – connections that reflect important formal parallels between the basic rules of theoretical and practical philosophy.

There is, however, one key feature in Watkins's account that my expansion has skipped over so far, namely, the claim that Kant's general notion of law is to be understood in terms of *prescription* by a "proper authority," a point that Kant makes not only in ethics but also in theoretical contexts, as in the first *Critique*'s statement that "categories prescribe laws a priori to appearances" (B163, cited at Watkins 2014a: 476; cf. *Prol* 4:320, "understanding does not draw its (a priori) laws from nature but prescribes"). Part, but only part, of the background for talk like this lies in the fact that the literal meaning of "law," long before Kant, was tied to practical contexts of a legal or religious nature, and so the common

[14] Cited in Watkins 2014a: 475, from Stark 2004: 53. [15] See Ameriks 2012: chapter 3.

[16] See Watkins 2014a: 479, and cf. Hebbeler 2015, on Kantian transcendental necessities as ungrounded fundamental bases for lawful universal derivations.

modern theoretical notion of laws of nature could at first seem like an inappropriate projection – unless it too was explained, as it often was, as an indirect way of expressing how a commanding authority, such as God, prescribes these laws (cf. Watkins 2014a: 477). As Michela Massimi has shown in her work, this tradition plays a very important role already in Kant's early writings, which try to provide a more satisfying explanation than Newton's specific way of thinking about God's prescribing nature's laws, for that way appears to be infected with considerable arbitrariness, dogmatism, and theological peculiarities.[17]

There is, moreover, a *general theoretical* impetus for talk about something like prescribing that can be found in broadly acknowledged points that go back before Newton and were stressed by Descartes. First, there is the point that, however fundamental the most basic laws of modern physics are, they (like many traditional moral commands as well) do not appear to be analytic or intrinsically necessary. Hence it is understandable that these lawful starting points – for example, the permanence of substance – often have been pictured as dependent on something like the selective power of a highest being. A second point is that even the best laws of modern physics traditionally presume an independent set of original material conditions on which the laws operate. Furthermore, even after modern philosophers worked out all that could be understandably deduced from these starting points, there still appeared to be a "superadded" excess of contingent order throughout the universe, especially in the organic domain.[18] As Kant famously remarks, even a "Newton" (*CJ* 5:400) of our age – in contrast to some new genius in the not to be counted on future – supposedly could never explain the simplest blade of grass, let alone the aesthetic appreciation of a sunset, by a deduction from the data and laws of our most basic exact sciences (which Kant's *Critiques* still limit to mechanics).

It thus was not far-fetched to regard the universal and surprisingly intricate organization of the world, revealed in more and more astonishing detail in the Newtonian era, as not a matter of necessity in the narrow logical sense, or even – contrary to Spinozists and Leibnizeans – as convincingly explicable to us in terms of any appeal to mere metaphysical notions such as sufficient reason and perfection, or any genuinely

[17] Massimi 2014. The three Kantian improvements that she explains can be matched up, I believe, with the interconnection of the three key concepts of universality, necessity, and law.

[18] On responses to the excess of order found in the world, one can compare the reaction of Cotes and Newton to the beauty of the world's diversified order (Massimi 2014: 494) with Kant's concern with teleology throughout his work.

foreseeable scientific theory of everything. Even when we add the many scientific developments that go beyond eighteenth-century mechanics, the best attempts at global accounts still leave at least two fundamental and unaccounted for contingent factors: the plurality of basic laws of nature and the brute givenness of its original conditions. These two fundamental factors, even if together they might after all *in principle* somehow explain all in the "excess" that *follows* upon them, cannot account for each other or their own shape and existence. From the beginning of his career, Kant was very aware of these factors, and it is notable that they concern limitations that arise even apart from the specific difficulties that his doctrine of transcendental idealism raises for anyone who would claim that we could ever, even in principle, achieve a *complete* explanation of existence, that is, one that does not already presume these factors. At this point, it seems that the ultimate alternative is either to accept the factors as surd elements or, in order to do justice to reason's constant demand for some kind of explanation of any facts, to find a way of speaking convincingly of some authority that prescribes at least something about them.

These familiar general points provide an understandable *precondition* – but only a precondition – for insisting on speaking of the fundamental *known* principles of our situation – if they are to be accounted for at all – as ultimately dependent on *basic transcendental truths* that can be regarded as prescribed in at least the minimal negative sense of not being known as in any way themselves *derivable from given contingent* truths, for these are all, *ex hypothesi*, dependent on the original factors. The hard problem for Kantians arises only right after this point, when pressure is felt to go beyond this negative claim and to add some rational *positive* content to the notion of prescription that does not appear arbitrary. It is often supposed that the obvious next step for a Kantian is simply to say that if it is not God or something external in itself that does the prescribing, then it must be the *human* mind that prescribes the most basic truths by somehow simply *creating* them out of its own resources. But taken literally, this kind of talk is a nonstarter. Given that Kant begins with the thought that the human mind is finite and receptive, he can hardly believe that we actually *make* the fundamental sensible structures of space and time. He surely knows that, taken literally, making itself is a spatiotemporal process (so it cannot itself be an explanation of the processes as such), and the option of a nonspatiotemporal quasi-making, a literal creation *ex nihilo*, is also excluded for us, given our finitude. Furthermore, Kant holds that reason's tools, the fundamental structures of the understanding (which are essential to any transcendental truths, theoretical or practical), the categories, like the fundamental principles of morality, have an absolutely necessary status that goes beyond us and so cannot be a matter of anything that is literally made by any beings that are limited like human minds.

There is, fortunately, an alternative to the nonstarter strategy of understanding transcendental "prescribing" in terms of a literal process of making. The alternative is to characterize the aspect of mind that in general is relevant here as nothing tied down to our limited human situation, but as simply, at first, the expression of the positive general *capacity* of using reason, in its distinctive highest sense, by *any* kind of being with reason.[19] This sense picks up on the literal meaning of "prescribe" as simply a matter of a kind of "writing" ahead of time, that is, a determining a priori. Hence it can be understood in terms of an appreciation for necessity as such, that is, the capacity to apprehend modality, and thus to understand not only logic but also all the putative necessary truths of mathematical nature, metaphysics, morality, and mind as such.[20] To say that reason specifically prescribes is primarily to contrast its *demanding* action with the asserting of truths in a nondemanding sense, what Kant calls mere "rules of skill or counsels" (*G* 4:417). That is, for Kant what ultimately matters in prescribing is not mere spontaneity or action as such, but reason's specific kind of action. What is distinctive for him about the faculty of reason as such – which is *not* the same as mere rationality – is simply its special capacity for formulating, endorsing, and thereby legislating claims that are expressed as having some kind of strictly necessary and therefore demanding, rather than optional, status. Because this kind of demanding status is not an explicable feature of mere given facts, let alone nature in itself, and because it is a feature linked directly to powers of intelligibility, it can make sense to say that laws that are prescribed are laws

[19] Cf. Pollok 2014: 510, 525, 530, who characterizes reason's prescribing in terms of an "acknowledgement" of the pure understanding's principles as "pure or universal laws of nature" (*Prol* 4:320), and also cites, at 515, *Metaphysik Pölitz* 1790/1, 28:575, which links "transcendental form" with the (epistemic) "act of determining," whereas "transcendental matter" is the "determinable."

[20] Kant does suggest (*CJ* §76) that modal distinctions, such as between actuality and possibility, would not be of concern for a being that, unlike us, has intellectual intuition. I presume, however, that since Kant would allow such a being to be all knowing, it is incoherent to say that it could not realize, as even we can, that, if it exists at all, it is true that it necessarily exists. So presumably Kant must be making a more moderate point, perhaps that reliance on *trying to find out* whether some truth is necessary rather than contingent is not an issue for a higher intellect. Similarly, Kant's characterization of the intuitive intellect's mind as "nondiscursive" is best understood as meaning just that this mind does not *need* to resort to the extensive temporal features that characterize our use of discursivity, and this does not make the truths that we need to understand through a discursive process remain altogether unknowable, even "instantaneously" for the higher intellect.

of reason, and in that sense of mind, albeit not in any sense that is limited to the contingencies of individual experience and psychology.[21] This point is consistent with still saying that much of what the laws of reason are *about*, and how the structures of reason are filled with empirical content, is precisely not something that is internal to mere reason as such. As Kant puts it in the *Prolegomena*, what specifically is connected to the mind, as reason, is not empirical content as such but appreciation for the lawfulness of the necessity in the truths governing that content (*Prol* 4:296, cited at Watkins 2014a: 485).[22] Or more exactly: "Lawfulness" as such just is necessity in the strict sense, which is the sense that Kant holds cannot be justified by any mere exposure to uniform sensory data or other contingent factors such as ultimately arbitrary social or supernatural forces. Kant also holds that, as transcendental, this lawfulness is strictly universal in a global sense because, as the Transcendental Deduction argues, given the a priori unity of space and time, the intellect's categories can "prescribe the law to nature" (B165) for "whatever objects may present themselves" (B159) to us.[23]

Understanding-prescribed lawfulness in this noncreationist sense can help in appreciating what Kant means in saying that a law of reason is not just something that happens to apply *to* a being, as a rule of inheritance may, unbeknownst, in fact apply to a young child, but is to be understood as prescribed or "self-determined" *by* reason to itself ("giving the law to itself," *G* 4:431). When a mature being with reason appreciates the *necessary* applicability of a law to its situation, it recognizes it as precisely the kind of principle that would not be appropriate *in that way* if it were presented as *warranted* simply through appeal to a faculty lower than reason, such as mere sensibility or imagination, and thus determined by contingencies. I have argued elsewhere that this point is crucial to understanding Kant's much misunderstood *Groundwork* argument for the basic "principle of autonomy" (*G* 4:433), that is, our moral self-determination by reason.[24] Many readers, especially in the Anglophone tradition, have been tempted to suppose that, for Kant, autonomy, like familiar forms of making, is literally some kind of extra force attaching to ordinary human beings *as such* (either individually or as a species), that is, as human and finite

[21] See the contrast drawn by Pollok 2014: 522, in a theoretical context, between the understanding's strict universal "legislation" and "an inventory of behavioral data."

[22] See also Pollok 2014: 521, who notes the contrast Kant makes, at *Prol* 4:318, between the plurality of experiential laws and lawfulness as such. On Kant's practical notion of a unique "*allgemeingesetzgebenden*" will (*G* 4:432), see Ameriks 2016.

[23] See Ameriks 1978, and Friedman 2014: 537. [24] See Ameriks 2016.

rule-givers. But a close examination of the arguments in the text makes clear that the autonomy that Kant is claiming is most basically a *normative* characterization[25] of *reason's principle* as having the nonderivative status of being *necessarily* lawful and therefore as related to a more than *merely* human faculty, one that cannot be dependent on any lower faculties (which would make it as such literally heteronomous, a *faculty* taking a law from elsewhere).

Readers have been surprised by the fact that Kant does not explicitly elaborate much on "a principle of autonomy" in later works. But there is no need for him to do so in these exact terms, since there are many ways to express his main idea here, which is simply the normative independence of reason from all "heteronomous" grounds in morality – hence his repeated contention that it is inconsistent to attempt to base a principle that must be normatively necessary for all beings with reason, namely, the moral law, on modes of access that are manifestly contingent, such as sensory givens or arbitrary theological posits. The faculty that Kant – quite unlike most Anglophone philosophers – calls will (*Wille*) as "practical reason" is nothing other than the faculty of being able to formulate and appreciate, and thus make it possible in principle to act on, the lawfulness of practical law as such. It is precisely the strict necessity of such practical law that contrasts to the loose universality of the multiplicity of conventional "laws" that by themselves reflect only "technical" and "pragmatic" rules. To say that reason prescribes morality *as* a law is therefore to stress that it does not and cannot take the law's status as such, its normative force, from elsewhere. Nor can it make the law out of nothing, or even out of some contingent matter concretely within a person, such as the mere desire of an individual human agent to preserve an identity. To say that this faculty prescribes is thus to say that it can *legislate* in the most fundamental law-providing sense, for it alone can comprehend and express law as what it distinctively is, namely, a strictly demanding rule, something categorically necessary rather than optional, contingent, fulfilling one accidental need among others.

The main consequence of these points is that, if, for Kant, reason's prescriptive power is, even in this practical context, still to be understood in terms of the appreciation of a special kind of normativity rather than in any terms of any literal "making" (let alone of an arbitrary or subjective sort), then this should be all the more clearly true for contexts in which

[25] Although it is not only that for Kant, because its relevant assertion in our own situation also involves a reference to the causal power of being able to *respond* absolutely freely to the moral law.

Kant's talk of reason's prescribing has nothing at all to do with what he calls a faculty of will. Such a position admittedly does nothing to defeat the skeptic who denies that we have any acquaintance with substantive necessity. But rather than assume that Kant is trying to meet such a skeptic, one can understand him as simply saying: suppose that we do allow that there is some acknowledged necessity (see *CPrR* 5:56), such as some basic fit of mathematics and the physical order to our knowledge, as well as in the absolute bindingness of morality, then what kind of *faculty* alone (and metaphysics) can we at least noninconsistently align with this perception?

A noteworthy feature of this account so far is that, although it insists on linking the notions of knowable necessity and mind in some sense, the account does not yet say anything about transcendental idealism, and so it is an extra question whether and where the lawfulness that reason prescribes needs to be characterized as a "merely ideal" lawfulness. Here it is significant that, because the moral law as such is nonspatiotemporal, and because spatiotemporality is essential to all of the *Critique*'s arguments for transcendental idealism, there is no ground for regarding *that* law – any more than necessity in general – as itself transcendentally ideal. Hence, insofar as Kant regards this law to be prescribed by reason as well, it cannot be the case that, for Kant, the mere notion of prescription must already bring with it an assertion of idealism, let alone subjectivism. Readers may get confused on this matter because it is true that Kant rejects numerous alternative accounts of morality that happen to be aligned with forms of transcendental realism. But this fact does not mean that a *general* antecedent rejection of realism is required for the mere understanding of a law's necessity. Where transcendental idealism comes in, in this part of Kant's philosophy, is simply in his view that the belief in our free ability to *respond* appropriately to the moral law would supposedly be undermined if one were to make the common mistake – which transcendental idealism excludes – of supposing that accepting a universal spatiotemporal scope for the deterministic laws of modern science would *immediately* rule out the possibility of any free choice on our part.

A new problem arises here, however, for the close connection just drawn between reason, morality, and strong necessity may now seem to make it difficult to understand Kant's talk of prescribing in the different context of the laws of *nature*, or at least the transcendental principles underlying them. This is because, even though Kant regards these laws as necessary in the strict rather than loose sense, their ultimate metaphysical status, unlike that of the moral law, is precisely not absolutely necessary. This point is reinforced further by the doctrine of transcendental idealism, which implies that the laws are not only logically contingent but also metaphysically conditioned by a higher level of reality, namely, things in themselves. The principles of the first *Critique*'s Transcendental Analytic are expressed and proven only for the context of spatiotemporal experience. And the basic sensible forms of that experience, and not merely our actual awareness of them, are regarded by the Critical Kant – in contrast to Newton and others – as not metaphysically absolute (even if they are also not derivative, as others had proposed, on the relations of empirical givens).

It is at this point that the fifth basic distinction, concerning constitutive versus regulative necessity, which earlier was noted only in passing, becomes relevant. It is true that Kant's spatiotemporal transcendental principles are not literally determined by mere reason, but rather by an a priori combination of the categorical forms of the pure understanding with our forms of pure sensibility. But even though the ultimate contingency of the latter forms entails that the principles themselves have a kind of contingent status, these principles are not simply loose statements of general facts but have the special status of being *transcendental constitutive* conditions; that is, they hold determinatively for *all possible* experience of our kind. This means that they must also have a kind of strong normative status, for supposedly they are *individually* essential in *justifying* the only kind of concrete theoretical knowledge that we can have. So, although, in contrast to what may be seen from a "God's eye view" they have a deep metaphysical contingency and, given transcendental idealism, even mask the fundamental features of things in themselves (including our selves), it remains true that they are each determining principles that our intellect cannot do without if it is to have theoretical knowledge at all; this is what it means to say they are "constitutive."

Here a final and very slight adjustment to Watkins's analysis is relevant, for he suggests a close connection between reason's lawful power, in Kant's constitutive sense, and the specific thought of the "Copernican Revolution" (Watkins 2014a: 490). There are many different ideas that interpreters have linked to Kant's vague reference to the "first thought" ("*ersten Gedanken*") of Copernicus (Bxvi), but insofar as many readers still tend to link this thought *immediately* to Kant's Critical idealism, and *they* may not realize that this idealism rests on arguments for the ideality of space and time that do *not* directly concern our intellect as such (and hence that ideality is derived already in the Dissertation in the section on sensibility rather than intellect), it seems preferable to make any general characterization of what Kant simply means by the prescriptive force of the intellect alone independent of any talk about Copernicus. It is true that the modern focus on scientific theorizing, in contrast to mere empirical data collecting, is very much on Kant's mind, but this general empirical contrast of the use of the intellect, as opposed to mere sense, is not sufficient to capture what Kant specifically means by the authority of intellect in the capacity of reason to be law-prescribing. Nor, for the reasons just given, is it appropriate to leave open the suggestion that in general there is an immediate connection, for Kant, between the notion of prescribing and transcendental idealism.

Along this line one can go on to show that there is a coherent positive way to understand Kant's talk of "our" prescribing rules even to nature, ways that do not rest on either subjectivist or idealistic doctrines – even though in fact the spatiotemporal *content* covered by these rules does need, on separate grounds, to be given an idealist interpretation. And that can provide a strictly necessary understanding of these rules, even while avoiding dogmatism or claims about what holds for all beings with reason. The principle of the Second Analogy, for example, is admittedly not the only way that causality can ever be specified for Kant, but it is meant as expressing a schematized transcendental necessity that is deeper than any specific Newtonian principles (see *Prol* 4:319, cited at Watkins 2014a: 483). For these are affirmed only for this particular world, and not for all worlds of "possible experience" of a spatiotemporal kind in which epistemic beings that are still basically like us could work out some other knowledge of nature. The Analogy expresses a transcendental law of nature, one that, in contrast to any Newtonian principle, is broad enough to be *essential* to satisfying *our* kind of intellect's most basic need *as* a possibly successful intellect, that is, a being with a theoretical mind that can make objective claims at all concerning what is in space and time.

It is true that Kant concedes and even stresses that this particular "need" is contingent and secondary to our moral calling. For, simply as human beings, we do not all *have* to engage with the theoretical principles of philosophical knowledge (let alone a specific school of physics and geometry), whereas morality, he holds, has a principle that we each cannot reject without denying our "ownmost" self. And yet, despite this second-place status with respect to our basic needs, it is also true that in our common life the transcendental laws of nature, unlike specific theories of physics (or even lower practical principles, such as the principle of prudence) are not ever subject to being corrected or overridden, because their transcendental status signifies for Kant that there can never be another way *for us* to characterize the nature that we experience. And since even the mere empirically oriented but strictly universal transcendental lawfulness of these principles cannot be accounted for in terms of any given and contingent data, but must be understood as the intellect's constitutively demanding conditions upon our empirical knowledge, it is not misleading for Kant to speak here too of "our mind" as "prescribing" these principles as laws, even if this is due to what he tends to call our "pure understanding" rather than reason in its very highest sense, which concerns unconditional truths. These principles do, after all, still express a highest substantive form of necessity that we can theoretically affirm. For, even if they cannot be claimed to satisfy reason as such (see the restriction implicit in Kant's first Critical sentence: "in one species," Aiii) through a grasp of what was called the traditional (3b) sense of metaphysical necessity, which holds for all possible worlds, they still hold strictly and strongly in the Kantian (3c) sense of transcendental necessity, which covers the experiential world of all beings subject to our conditioned but universal forms of sensibility.

In sum, Kant's statements about our mind as law-prescribing claim neither too much, insofar as they back off from theoretically determining unconditioned reality in itself, nor too little, insofar as their admittedly subjective-sounding language actually reinforces rather than undercuts the strict and substantive objectivity that they intend.

CHAPTER 3

Imperfect Knowledge of Nature

Kant, Hume, and Laws of Nature

Paul Guyer

3.1 Hume, Kant, and the Problem of Induction

In the *Prolegomena to Any Future Metaphysics*, Kant asserted that Hume's conclusion that "a single but important concept in metaphysics, namely that of the *connection of cause and effect*, . . . is really nothing but a bastard of the imagination" by means of which "reason completely and fully deceives herself . . . , falsely taking it for her own child" and passing "off . . . subjective necessity (i.e., habit) for an objective necessity (from insight)" is "premature and erroneous" (*Prol*, Preface, 4:257–258). Kant did not think that Hume denied the importance of the concept of causation in our thought, but rather that he denied that the concept is a priori: "The question was not, whether the concept of cause is right, useful, and, with respect to all cognition of nature, indispensable; it was rather whether it is thought through reason *a priori*" (*Prol* 4:258–259). Clearly Kant intended to establish, in opposition to Hume, that the concept of causation is a priori, and indeed but one of a system of concepts, namely the pure concepts of the understanding that are not only "indispensable with respect to all cognition of nature" but are all a priori (*Prol* 4:260).

Kant knew Hume primarily from the *Enquiry Concerning Human Understanding*[1] rather than from the earlier *Treatise of Human Nature*.[2] Yet it is in the *Treatise* that Hume foregrounds his concern that there is a problem with the origin of the *concept* of causation, while the *Enquiry* focuses on the question of whether our knowledge of actual *causal connections* or, as we would say, particular *causal laws*, is "attained by reasonings *a priori*," or only "arises entirely from experience" (*EHU*, section 4, part I: 25). In particular, the *Enquiry* foregrounds Hume's famous argument that

[1] The first *Enquiry* (or *Philosophical Essays Concerning Human Understanding*, as the text was originally called), was translated in Hume 1755.

[2] The *Treatise* was first published in its entirety in German only in 1790–1792, as David Hume, *Über die menschliche Natur* (Hume 1790–1792).

induction must be based on the assumption that "the future will be conformable to the past" (*EHU*, section 4, part 2:31) but that this assumption is not logically necessary nor can it, on pain of circularity, be derived from experience, so that the practice of induction, indispensable as it is to us, and "though none but a fool or madman will ever pretend to dispute the authority of experience" (31), has no rational foundation, but is only a "CUSTOM or HABIT" (*EHU*, section 5, part 1: 37) grounded in human psychology. Yet Kant never directly discussed this argument, a fortiori never directly confuted it.

Several authors have attempted to locate a rebuttal of Hume's worry about induction in the Introduction to the *Critique of the Power of Judgment.*[3] Eric Watkins, although he thinks that Kant was attempting to offer an alternative to Hume's approach to causation rather than a refutation of it based on one or more shared premises, has suggested that Kant "might have thought" that an answer to Hume's worry about whether the future must be conformable to the past could be grounded in his demonstration of the existence of permanent and immutable substance.[4] But I find no evidence that Kant was thinking about Hume's worry about induction in either of these contexts. Perhaps this is because Kant's own position is that knowledge of particular causal laws, or laws of nature, is possible only as part of knowledge of a complete *system* of such laws. Yet the completeness of such a system always remains only a regulative ideal for us and places a more genuine restriction on our knowledge of laws of nature than does Hume's worry about induction. On Hume's own account, his worry about induction is academic, that is, an academic exercise to be forgotten as soon as he steps out for dinner and a game of billiards with his friends. Everyday human psychology entirely redeems us from this academic worry.[5] But Kant's concern about the merely regulative ideal of the completeness of empirical knowledge and of everything that would depend on such completeness is no merely academic exercise, but a substantive constraint on the conduct of enquiry. It is a manifestation of

[3] E.g., Floyd 1998. [4] Watkins 2005.

[5] This is hardly to say that all of Hume's concerns about causation are merely academic. On the contrary, his insistence that we can make causal inferences only of objects of which we have had repeated experience is Hume's great cudgel against causal arguments for the existence of God, in both the forms that Kant would call the cosmological argument, from the contingent existence of the world to God as its necessary cause, and the physicotheological argument, the argument from design, the argument from the intelligent design of the world to the existence of its intelligent designer. But the argument about induction is never part of Hume's critique of arguments for the existence of God: he never argues that in the past gods were always necessary to create worlds, but might not be so in the future (thanks to Graciela de Pierris for pushing me to address this point).

epistemological modesty about our claims to empirical knowledge of particular laws of nature that complements the epistemological arrogance of Kant's claim for the completeness of his table of a priori categories and most general synthetic a priori principles of empirical knowledge. Perhaps we can allow the speculation that Kant did not draw attention to Hume's worry about induction because in the face of Enlightenment optimism, he did not want to draw too much attention to his own strictures on the possibility of empirical knowledge of the laws of nature.

I have addressed how Kant does attempt to obviate Hume's concerns about the origins of our *concept* of causation through the schematism of the category of the relation of ground and consequence and the argument of the Second Analogy of Experience elsewhere.[6] Here I will focus on the strictures on our knowledge of particular causal laws of nature that are inherent in Kant's conception of the systematicity of the concepts and laws of nature as a merely regulative ideal, which we can regard as Kant's substitute for, rather than rebuttal of, Hume's worry about induction.

3.2 Kant on the Necessity for Knowledge of Causal Laws

Kant's argument about objective time-determination in the Second Analogy of Experience requires that we have knowledge of particular causal laws. For we can determine that a change in our own perceptions represents a change in states of affairs external to those perceptions and what change it represents only by subsuming the two states under some rule that entails that under the relevant circumstances one of those states must have preceded and one followed, which is precisely what a particular causal law does. To use Kant's famous contrast between the cases of an immobile house and a moving ship (A192–193/B237–238), I can judge that my succession of representations of the positions of a ship in a river represents a change in its position but that my succession of perceptions of the parts of a house does not represent any change in the house only because I can appeal to causal laws about rivers and ships that entail such a change in position under the relevant conditions. By contrast, in the case envisioned, there are no such laws entailing that the house has undergone any change.[7]

6 See Guyer 2003a.

7 Of course, as Arthur Schopenhauer pointed out, there is a causal explanation of the succession of my perceptions in the case of the house, too, namely, the changes in the position of my eyes as I move my head, and so on; see Schopenhauer 1813/2012, §23:83–84. The point that Schopenhauer misses is that Kant is arguing only that the laws relevant to the case of the ship entail precisely that the ship is moving, while the laws relevant to the case of the house entail only that the observer is moving, not that the change in perception of the house involves no causality anywhere.

Many things might be said about this argument, but here I want to focus on one point. Hume famously distinguished the question "How experience gives rise" to the *general* principle "of the necessity of a cause to every new production," or that every event must have some cause, and the question "Why we conclude, that such particular causes must have such particular effects," or how we know *particular* causal laws, and then said it might be "convenient to sink" the former question in the latter (*THN*, I.III.iv.9:58). Almost as famously, Lewis White Beck took Kant to distinguish the same two issues, and takes the Second Analogy to be aimed at proving the more general principle, as his translation of Kant puts it, that "Everything that happens, that is, begins to be, presupposes something upon which it follows by rule." But Beck also argues that Kant infers this *from* his explanation of how "Events can be distinguished from objective enduring states of affairs."[8] And this is correct: what Kant is doing is proving the general principle that every event (that we can know to be an event and the particular event that it is) must be entailed by some particular causal law(s) or other. Every event must have some particular cause(s) because it cannot be known to be an event and to be the event that it is without such laws.[9]

But this is hardly to say that in practice actual knowledge of all the particular causal laws of nature precedes our knowledge of the general principle that every event must have some cause, that is, follow some other event in accordance with a rule. The argument induces our knowledge of the general principle not from any antecedent *knowledge* of particular causal laws, but only from its account of the *role* of particular causal laws in our knowledge of events. Neither does Kant suppose the reverse, that our knowledge of particular causal laws somehow follows directly from our knowledge of the general principle that every event has a cause. He has already made this clear in the Transcendental Deduction, for instance in the second edition when he stated that "Particular laws, because they concern empirically determined categories, **cannot** be **completely derived** from the categories, although they all stand under them. Experience must be added in order to come to know particular laws **at all**" (B165). Kant's inference that every event must have some cause from the explanation that we cannot determine that an event and what event has taken place without

[8] See Beck 1967: 135. Beck puts this point by calling the general principle "K," the second principle "P," and by saying "P implies K, by the arguments of the Second Analogy."

[9] I have made this point in my discussions of Kant on causation since Guyer 1987: 241–259. Michael Friedman also argued that Kant's argument is meant to show that we need to know particular causal laws in Friedman 1992b.

appeal to particular causal laws has not itself given us or been meant to give us an account of how we know particular causal laws.

Kant's remark that we have to "add" experience to the categories to know particular laws might suggest that we simply have to add observations to the categories for that end. But that seems implausible precisely in light of the stricture on immediate perception from which he begins the Second Analogy. Rather, Kant's view about the basis for our knowledge of particular causal laws seems adumbrated by his remarks about the systematicity of knowledge in the Appendix to the Transcendental Dialectic of the first *Critique* and the Introduction to the third, although neither of these texts mentions Hume nor directly addresses his worry about the rationality of induction. What Kant argues in these passages is that we can know particular causal laws only as part of a system of such laws, which implies that the limits on our knowledge of the necessity and completeness of any such system will also be limits on our knowledge of the particular laws that we know through it. My contention is that these are substantive limits on our knowledge of nature rather than the academic worry raised by Hume's argument about induction.

3.3 Kant on the Regulative Idea of Systematicity

Kant's opening remarks in the Appendix to the Transcendental Dialectic of the *Critique of Pure Reason* might suggest that systematicity is an interest of reason that reason seeks to realize for its own satisfaction in a manifold of concepts (or laws) already given to it as such by the understanding: "Reason really has as object only the understanding and its purposive application, and just as the understanding unites the manifold into an object through concepts, so reason on its side unites the manifold of concepts through ideas by positing a certain collective unity as the goal of the understanding's actions," Kant says in the Appendix (A643–644/B671–672). Similarly, he says in a note that "Just as the senses are related to the understanding, so is the understanding related to reason. The appearances of the first acquire the unity of the understanding in the second through concepts, and concepts acquire the unity of reason in the third faculty through ideas" (R 5553, 18:221–222; *NF*:239). These passages may make it sound as if the special kind of unity among concepts that reason can find is its own desideratum, not necessary for the understanding to do its own proper work.

A few pages later in the Appendix, however, Kant says, "The hypothetical use of reason is ... directed at the systematic unity of the understanding's cognitions, which, however, is the **touchstone of truth** for its rules" (A647/B675) and that "For the law of reason to seek unity is necessary, since without it we would have no reason, and without that, no coherent use of the understanding, and, lacking that, no sufficient mark of empirical truth; thus in regard to the latter we simply have to presuppose the systematic unity of nature as objectively valid and necessary" (A651/B679). If the last quotation stopped after the phrase "no coherent use of the understanding," we might be able to distinguish the use of the understanding in general from some especially systematic use of it and think of the latter as a desideratum of reason, not of the understanding itself. But the sentence goes on to state that the kind of systematic unity of nature found only by reason is in fact necessary in order to have a *sufficient* mark of empirical truth. I will take this to be Kant's considered position, namely, that while the understanding, through general and transcendental logic, supplies certain *necessary* conditions of empirical truth, thus of the truth of empirical laws of nature and the objective validity of the concepts that express them, in the form of the basic laws of logic and the basic categories for concepts of objects, only reason, through its requirement of systematic unity, can specify what is necessary for *sufficient* conditions of empirical truth, or the truth of empirical laws of nature. Of course it will be the work of judgment, ultimately reflecting judgment, to satisfy the requirements for systematicity laid out by reason,[10] although perhaps *inferring* one judgment from others, as we will see, is made possible by systematicity, and can also be considered the work of reason.

The first question to consider then is precisely what sufficient condition for empirical truth does systematic unity add. To answer this question, of course, we must first have Kant's conception of systematic unity before us. In the Appendix, Kant defines systematic unity by the three principles of homogeneity, specificity, and affinity:

> reason ... prepares the field for the understanding 1. by a principle of **sameness of kind** [*Gleichartigkeit*] in the manifold under higher genera, 2. by a principle of the **variety** of what is the same in kind under lower species; and in order to complete the systematic unity it adds 3. still another law of the **affinity** of all concepts, which offers a continuous transition from every species to every other through a graduated increase of varieties. We can call these the principles of the **homogeneity**, **specification** and **continuity** of forms. (A657–658/B685–686)

[10] See Guyer 1990.

In accordance with the principle of homogeneity, we may suppose that "the nature of things themselves offers material for the unity of reason, and the apparently infinite variety should not restrain us from conjecturing behind it a unity of fundamental properties, from which their manifoldness can be derived only through repeated determination" (A652/B680). That is, this principle posits that the variety of natural phenomena can always be subsumed under higher-order, unifying concepts, in principle up to the limit of a single fundamental power underlying all natural phenomena. In accordance with the principle of specification or specificity, we may suppose, conversely, that we can always divide higher-order classifications into more specific ones, subsume species under genera and subspecies under species, but without ever bridging the insuperable gulf between concepts, that is, classificatory generalities comprised of a finite number of marks, and individuals, that is, particulars comprised of an indefinite or potentially infinite number of properties:

> reason demands in its entire extension that no species be regarded as in itself the lowest; for since each species is always a concept that contains within itself only what is common to different things, this concept cannot be thoroughly determined, hence it cannot be related to an individual, consequently, it must at every time contain other concepts, i.e., subspecies, under itself (A655–656/B683–684).

Finally, Kant states about the principle of continuity or affinity of forms that it "arises by uniting the first two, according as one has completed the systematic connection in the idea by ascending to higher genera, as well as descending to lower species; for then all manifolds are akin to one another, because they are all collectively descended, through every degree of extended determination, from a single highest genus" (A658/B686). His thought here seems to be that concepts within a system must be so organized that one could reach any one from any other, by ascending from any one lower-order concept to that higher-order concept, which is high enough so that from it a different descent to the other can be found. For example, to understand the relation between, say, Yorkies and St. Bernards we have to be able to ascend to their common membership in *Canis familiaris* down through their types, lapdogs and working dogs (as recognized, e.g., by the American Kennel Club), to their specific breeds, although not all the way down to any particular Yorkie and St. Bernard. To understand the relation between dogs and bears, we have to work our way up to the families *Canidae* and *Ursidae* and beyond that to the order *Carnivora*, and so on. We will see the significance of this shortly.

But before we do, one comment is in order. Kant's talk of genera, species, and subspecies and my example with dogs and bears may make it sound as if Kant is talking exclusively about the structure of taxonomies of types of *substances*, here organisms, but perhaps in other cases elements, such as in the periodic table, or minerals, such as gems, and so on. However, in the characterization of the principle of homogeneity that was cited, Kant spoke of fundamental *properties* rather than kinds of *beings*, thus it may seem that his scheme is meant to apply to properties like walking and swimming, which can be subsumed under higher-order concepts like locomotion as well as broken down further, say into bipedal, quadrupedal, and so on, in addition to, or even instead of kinds of beings, for example, organisms, such as humans, other vertebrates, insects, and so on. In fact, Kant's own most extended example of a systematic classification applies to *patterns of motion* as much as to *kinds of objects*. We will refer to this example again in the sequel, so it will help to have it before us now:

> If, e.g., the course of the planets is given to us as circular through a (still not fully corrected) experience, and we find variations, then we suppose these variations to consist in an orbit that can deviate from the circle through each of an infinity of intermediate degrees according to constant laws; i.e., we suppose that the movements of the planets that are not a circle will more or less approximate to its properties, and then we come upon the ellipse. The comets show an even greater variety in their paths, since (as far as observation reaches) they do not ever return in a circle; yet we guess at a parabolic course for them, since it is still akin to the ellipse and, if the major axis of the latter is very long, it cannot be distinguished from it in all our observations. Thus under the guidance of those principles we come to a unity of genera in the forms of these paths, but thereby also to unity in the cause of all the laws of this motion (gravitation); from there we extend our conquests, seeking to explain all variations and deviations from those rules on the basis of the same principle; finally we add on even more than experience can ever confirm, namely in accordance with the rules of affinity, even conceiving hyperbolical paths for comets in which these bodies leave our solar system entirely and, going from sun to sun, unite in their course the most remote parts of a world system, which for us is unbounded yet connected through one and the same moving force. (A662–663/B690–691)

This example illustrates numerous points. First, it shows how Kant thinks that the systematic unity of concepts relates concepts to one another: finding that the concept of circle does not quite work for the orbits of the planets in our solar system, we ascend to a more general class of closed curves that also includes ellipses, then descend from the concept of the ellipse down to the more specific kind of ellipse, as defined by the ratio between its major and minor axis, which best fits the observed motions of the planets. By discovering another path that does not quite fit the model of an ellipse, namely, the path of comets, we ascend back up to the general concept of closed curves, then descend down to the more specific concept of parabolic curves, and from there to the specific curve that fits our observations of comets; and so on. Second, the illustration makes clear that the model of systematic unity can be applied to patterns of motion or behavior as well as to types of objects; thus, it shows how the concepts of planets, comets, and so on can be subsumed under a higher-order concept of celestial bodies, and then no doubt also further subdivided, as into comets with different parabolic paths, but also how the different orbits themselves can be classified. Third, it suggests that the classification of types of behavior, in this case, orbital motions, is connected to *causal laws* and their types; thus, Kant explicitly states that the systematic relations among the different kinds of orbits allows us to see them as different effects of a common cause, namely, gravitation, but we can also see the different orbits as themselves different types of causes, having systematically different effects on neighboring bodies in accordance with the more general law of gravitation, and so on. Finally, Kant's illustration also suggests how the systematic unity of classificatory and explanatory *concepts* plays a crucial role in our cognition of the world or universe itself as a systematic *object* or *system of objects*: the law-governed trajectory of a comet through the solar systems is part of what makes the several solar systems comprise parts of a single galaxy, and so on. We might think of this last point as a massive extension of the thought underlying Kant's Third Analogy of Experience. Just as we have to recognize dynamic laws of interaction between any two bodies in space that we can judge to exist simultaneously, so can we recognize any number of objects as comprising a genuine community or whole only on the basis of a system of causal laws systematically connecting them all.[11] It is in this way that a system of concepts and a system of objects go hand in hand.

We can now turn at last to Kant's claim that the systematic unity of reason is not merely a desideratum of reason that is, so to speak,

[11] Jeffrey Edwards argues that the Third Analogy problematizes Kant's distinction between purely transcendental and properly metaphysical principles, of the kind that should be proven only in the *Metaphysical Foundations of Natural Sciences*; see Edwards 2000: chapter 3. But precisely because the inference to a genuine community of cosmic matter depends on an assumption that it is possible to determine simultaneity throughout the universe (which is questioned by the Special Theory of Relativity), I would maintain that this inference illustrates the difference between the pure categories given entirely a priori and an actual system of knowledge of objects and laws, given only regulatively. See also Friedman 2013, pp. 563–80, especially pp. 578–9.

supererogatory from the point of view of the understanding, but is rather necessary to complete sufficient conditions for empirical truth itself. What could Kant mean by this claim?

One thing that Kant's example as well as several subsequent comments suggest is that his model of systematicity is meant to serve merely as a *heuristic*, as a *method for discovering* particular empirical truths, including the truth of particular empirical laws. In the paragraph following the illustration, he says that "even though" the principles of systematicity "contain mere ideas to be followed in the empirical use of reason, which reason can follow only asymptotically, as it were, i.e., merely by approximation . . . yet these principles, as synthetic propositions *a priori*, nevertheless have objective but indeterminate validity; and serve as a rule of possible experience, and can even be used with good success, as heuristic principles, in actually elaborating it" (A663/B691). And in the first draft of the Introduction to the third *Critique*, where Kant has reassigned the idea of systematicity from reason to the newly introduced faculty of reflective judgment, he says that this power

> proceeds with the given appearances, in order to bring them under empirical concepts of determinate natural things, not schematically, but **technically**, not as it were merely mechanically, like an instrument, but **artistically**, in accordance with the general but at the same time indeterminate principle of a purposive arrangement of nature in a system, as it were for the benefit of our power of judgment, in the suitability of its particular laws (about which understanding has nothing to say) for the possibility of experience as a system, without which presupposition we could not hope to find our way in a labyrinth of the multiplicity of possible empirical particular laws. (*CJ* 20:213–214)

The last clause of this passage suggests that the idea of systematic unity gives us a method to seek particular empirical laws rather than just looking around among endless possibilities in a way that is merely "arbitrary and blind" (*CJ* 20:214). The heuristic would be to construct concepts by filling in gaps or nodes in our system by ascent or descent, and then looking for empirical data to confirm the objective reality of such concepts. For example, if the perfect circle does not work as the pattern for planetary orbits, ascend one level of generality to the class of curves that includes ellipses as well as circles, then descend back from there to the ellipse, and from there to the particular kind of ellipse that you think will best fit the observational data – and maybe you will have found the right orbit. If it still does not fit, repeat the procedure.

The idea of a heuristic, however, seems to be the idea of something *subjective*, an approach that *we use* to organize our conduct of inquiry but not something that we need suppose to exist *objectively*, in the objects themselves. In other words, the idea of a heuristic would not seem to live up to Kant's claim that the idea of systematic unity is a touchstone or necessary condition for truth; it would seem to provide us with an organized method to search after truth, but not to be any part of what determines whether a particular hypothesis *actually is true*. This idea would also not seem to sit well with Kant's repeated claims in both the first and third critiques that the idea of systematic unity is a *transcendental* and not just a *logical* principle. For example, "In fact it cannot even be seen how there could be a logical principle of rational unity among rules unless a transcendental principle is presupposed, through which such a systematic unity, as pertaining to the object itself, is assumed *a priori* as necessary" (A650–651/B678–679; cf. A654/B682; A657/B685; A660/B688; *FI* 20:214–215; *CJ*, Introduction, Section V, 5:181–182). A logical principle would be one that posits merely a certain structure among our own concepts; a transcendental principle instead presupposes a corresponding structure in nature itself (though, of course, nature as appearance, not as a thing in itself). A purely heuristic principle that tells us that it is rational to conduct our search for truth in a specific way need not posit antecedently that nature or reality displays a structure analogous to that in which it is most rational for us to proceed in formulating and testing concepts, although no doubt it would be irrational for us to continue to employ this heuristic if we knew that there was something in the structure of nature itself that is incompatible with it. The rationality of using the heuristic would certainly not require that we know that nature *necessarily* conforms to it, as Kant's claim at A650–651/B578–579 suggests we must.

Instead, I suggest that Kant has something different in mind. What systematicity contributes to the empirical truth of particular laws of nature, beyond compliance with the formal requirements of general and transcendental logic and compatibility with actual observations, is the necessity of those laws. In the published version of the Introduction to the third *Critique*, Kant argues that particular laws of nature must be necessarily true, do not derive their necessity from the categories or most general principles of nature alone, and therefore must have some other source for their necessity, which he suggests lies in the membership in a system of laws. The key passage is the following:

> The understanding is of course in possession *a priori* of universal laws of nature, without which nature could not be an object of experience at all; but it still requires in addition a certain order of nature in its particular rules, which can only be known to it empirically and which from its point

> of view are contingent. These rules, without which there would be no progress from the general analogy of a possible experience at all to the particular, it must think as laws (i.e., as necessary), because otherwise they would not constitute an order of nature, even though it does not and never can cognize their necessity. Thus although it cannot determine anything *a priori* with regard to those (objects), it must yet, in order to investigate these empirical so-called laws, ground all reflection on nature on an *a priori* principle, the principle, namely, that in accordance with these laws a cognizable order of nature is possible – the sort of principle that is expressed in the following propositions: that there is in nature a subordination of genera and species that we can grasp; that the latter in turn converge in accordance with a common principle, so that a transition from one to the other and thereby to a higher genus is possible . . . etc. This agreement of nature with our faculty of cognition is presupposed *a priori* by the power of judgment on behalf of its reflection on nature in accordance with empirical laws, while at the same time the understanding recognizes it objectively as contingent, and only the power of judgment attributes it to nature as transcendental purposiveness . . . (*CJ*, Introduction, Section V, 5:184–185).

Kant recognizes that the "universal laws of nature," by which he presumably means the principles of pure understanding established in the first *Critique*, including especially the Analogies of Experience, to which he may be specifically alluding with the phrase "the general analogy of a possible experience at all," do not suffice to determine the necessity or necessary truth of any particular laws of nature. And this seems entirely correct, for as Eric Watkins has pointed out, the kind of interpretation of the Second Analogy in particular that I have defended shows how the sequence of states of affairs in a particular event must be *necessitated by* the relevant causal laws, or how it is necessary *relative* to those laws, but not why *those laws themselves* are necessarily true.[12] But while Watkins makes this point in criticism of my interpretation of the Second Analogy, I suggest that Kant's own recognition that he still owes us an explanation of our cognition of the necessity of particular laws of nature actually confirms that interpretation: the Second Analogy does indeed show us the role of particular causal laws in the cognition of events, thus the necessity of causal laws *for* time-determination, but it does not and at least from Kant's point of view in the third *Critique* was not intended to explain the de se necessity of the causal laws themselves. Only the position of particular laws within a system, thus the fact that within a system we can derive more particular laws from more general ones, can lend those laws

[12] See Watkins 2005: 211.

themselves a kind of necessity. To be sure, Kant does not say much more about *how* the derivation of such necessity works beyond his remark that within a system "transition" from one law to another is possible. However, we can at least suppose that by this he means that within a systematic arrangement of laws those at one level may be deduced from those at another level – perhaps typically those at a lower level from some at a higher level, but maybe also more general, thus higher laws from a multiplicity of laws at a lower level – and that this sort of transition is the closest we can come to the kind of rational deduction that would reveal the necessity of particular laws.

This line of thought explains why Kant is insistent that the principle of systematicity must be a transcendental and not merely a logical principle, assuming that a merely heuristic or methodological principle would count only as the latter. If we are to think that its position in a system is what makes any particular law of nature necessarily true, we must think that the system exists independently of our representation of it, particularly in the case in which we may suppose we know the particular law but not all of the system that is its context and supplies necessity: we can still suppose that the system makes it necessary even if we do not know the whole system, but we could not suppose this if we also supposed that no more system exists than what we ourselves currently happen to know. That might suffice to lend the appearance of necessity to some of the laws we take ourselves to know, but not to all.

I should add here that the contrast between logical and transcendental, or between internal and external, subjective or objective, that I am assuming here is what we might call the ordinary, empirical contrast and not Kant's transcendental idealist contrast between appearances and things in themselves. What I am suggesting is that Kant is arguing that for us humans to see particular causal laws as necessarily true requires that we must suppose them to be part of a complete system of such laws that obtains beyond our own partial representation or imagination of it, as it were outside our own individual or collective skins rather than merely within our individual or collective brains. This involves the empirical rather than transcendental contrast between inner and outer (see A373). It is independent of the core claim of transcendental idealism that the spatiotemporal form of all our representation of the world is a feature of the way we represent the world rather than of the world as it is in itself. But we can reconcile the transcendental status of the principle of systematicity, which essentially claims external reality for it, with the transcendental idealism of spatiotemporal form itself, in the same way in which Kant reconciles empirical realism and transcendental idealism in general. In the general case, we do that by supposing that things other than our own representations really do exist, although not in the spatiotemporal form in which we represent them (*Prol*, §13, note II, 4:289, and note III, 4:293). Similarly, we can reconcile what is in fact the transcendental principle of the empirical realism of systematicity with transcendental idealism by supposing that the things that really do exist independent of our representations of them also really do have systematic relations among their powers, although those powers do not really have the spatiotemporal forms that they appear to us to have and that are captured by our inescapably spatiotemporal representation of them through causal laws rather than through some less specific laws of ground and consequence relations.

We can take Kant himself to be signaling that the empirical realism of the systematicity of natural laws must and can be reconciled with transcendental idealism by means of his argument that "the particular empirical laws, in regard to that which is left undetermined in them" by *our* understanding "must be considered in terms of the sort of unity they would have if an understanding (even if not ours) had likewise given them for the sake of our faculty of cognition, in order to make possible a system of experience in accordance with particular laws of nature" (*CJ*, Introduction, Section IV, 5:180). Kant's key argument for transcendental idealism, of course, is the argument that only transcendental idealism can explain the possibility of synthetic a priori cognition: that the only thing a mind can know to be necessarily true is what in fact depends on that mind (see A26/B42; A47–49/B64–66; *Prol*, §13, note I, 4:287–288). Or at least this is the only way that *we* can understand necessity. Thus in the case in which we cannot explain necessity by dependence on our *own* minds, as we do when we attribute the necessity of spatiotemporal form and the most universal laws of nature to the structure of our own representation of objects, we at least are still disposed to attribute the necessity in question to dependence on *some* mind, some mind other than our own. (The attribution of necessity to a divine mind might seem to be in tension with a theological conception of God as absolutely free. But Kant notoriously argues that the appearance of possibility is due to the gap between concepts and intuitions for humans, while for God, with his intellectual intuition, there is no such gap, thus for that reason his thought is thought of a completely necessary world; see *CJ*, §§76–77. In this context, at least, Kant does not worry about divine freedom).

3.4 Kant on the Imperfect Knowledge of Nature

However, as soon as we follow our natural bent to attribute necessity to mind-dependence, even if the mind in question must be other than our own, we also undermine any claim to *know* that necessity, for the key to *knowledge* of necessity is precisely that it is a kind of *self-knowledge*. Then again, Kant's claim is not that we actually *know* the necessity of particular causal laws, but only that we have to have a way of *representing* them as necessary, or of thinking of them *as if* they were necessarily true. Indeed, Kant's ascription of the necessity of the particular causal laws of nature to a mind other than our own can be taken as a metaphor for his substantive rather than merely academic restriction on the certainty of our knowledge of the laws of nature.

Kant is clear that the attribution of systematicity in nature to a mind other than our own must be modified with an "as if." Thus, in the second half of the Appendix to the Transcendental Dialectic of the first *Critique*, discussing the psychological, cosmological, and theological ideas of pure reason, he says that the first connects "all appearances, actions, and receptivity of our mind to the guiding thread of inner experience **as if** the mind were a simple substance"; that in accordance with the second "we have to pursue the conditions of the inner as well as the outer appearances of nature through an investigation that will nowhere be completed, **as if** nature were infinite in itself," and then, most important for our present concern, that

> (in regard to theology) we have to consider everything that might ever belong to the context of possible experience **as if** this experience constituted an absolute unity, but one dependent through and through, and always still conditioned within the world of sense, yet at the same time **as if** the sum total of all appearances (the world of sense itself) had a single supreme and all-sufficient ground outside its range, namely an independent, original, and creative reason, as it were, in relation to which we direct every empirical use of **our** reason in its greatest extension **as if** the objects themselves had arisen from the original image of all reason. That means . . . it is not from a highest intelligence that we derive the order of the world and its systematic unity, but rather it is from the idea of a most wise cause that we take the rule that reason is best off using for its own satisfaction when it connects up causes and effects in the world. (A672–673/B700–701)

The last of these lines reminds us of Kant's original suggestion that systematicity is just an interest of reason "for its own satisfaction," not a necessary condition for understanding, which we would have thought to have left long behind. The preceding remark that we should "direct every empirical use of our reason . . . as if the objects themselves had arisen from the original image of all reason," that is, a divine rather than merely human reason, might fit better with the merely heuristic use of the idea of systematicity that Kant so often suggests rather than with the necessity-constituting role for it that he explicitly asserts at least in the third *Critique*. But his clear commitment to the "as if" status of our idea of a supreme intelligence has to be reconciled with the latter. This can best be done by adopting a complex attitude suggested by the relation between empirical realism and transcendental idealism: while at one level – call it the empirical level – we have to think of the systematic order of nature as existing independently from our always incomplete representation of it in order to think of it as lending necessity to particular empirical laws even when we are not yet in full possession of their systematic context, at another level – call it transcendental – we have to recognize that the very idea of the complete system and of its source in a supreme intelligence is in fact our own idea. Of course, this is emblematic of the complex dance that Kant always does with the idea of God, down to his dying days, as when he argues in the *Opus postumum* that we need to represent pure practical reason as emanating from God, but from a God who is not a substance but nothing other than our own idea.[13]

Kant thus signals that our knowledge of the necessity of natural laws is always imperfect – our knowledge of the whole body of such laws is always incomplete, and therefore our knowledge of the necessity of any particular law is always provisional – by qualifying our assertion of the existence of the systematicity on which such necessity depends and that of the intelligence on which the systematicity in turn depends with the phrase "as if." He similarly reminds us in the third *Critique* that "This agreement of nature with our faculty of cognition is presupposed *a priori* by the power of judgment on behalf of its reflection on nature in accordance with empirical laws, while at the same time the understanding recognizes it objectively as contingent" (*CJ*, Introduction, Section V, 5:185). Here again we have to take a complex attitude toward ourselves, at one level proceeding as if nature really is systematic independently of ourselves and thus even that its particular laws are necessary. But at another level we recognize – ultimately because we can have no objective knowledge of the actual existence of the supreme intelligence on which its systematicity would depend – that we

13 I have discussed Kant's treatment of the concept of God in the *Opus postumum* in Guyer 2000. For a related approach, see Förster 2000: chapter 5, 117–147.

can regard the existence of this systematicity itself only as contingent.[14] We might say that by Kant's own account the threatened contingency of causal laws themselves has been pushed back only one level: first we worried that even though particular events themselves may be necessary relative to the causal laws that cover them, those laws themselves are contingent. Then we sought to ameliorate the latter charge by including those causal laws within a system of laws that would lend a kind of necessity to its individual members in virtue of its internal relations of entailment. But, finally, we have to recognize that there is still a sense in which the existence of any such system as a whole must remain contingent because we can act only *as if* the supreme intelligence, on which it must rest, exists, and we cannot really *know* that it does. Thus Kant does not think we should worry about whether bread will *continue* to nourish or billiard balls *continue* to transmit their velocities in accordance with the laws of mechanics. But he does insist that we can never know the whole system on which rests our affirmation of the *necessity* of the particular laws governing such occurrences to be itself necessarily true. Kant is not worried that genuine laws of nature might actually change or cease to hold, but rather points out that our assertion of such laws is in the final analysis, in the terms of his anatomy of *Fürwahrhalten* or propositional attitudes in the first *Critique*'s Doctrine of Method, always ultimately opinion rather than genuine knowledge. Kant never explicitly says that, and "opinion" might be misleadingly negative in English. Maybe we should call our knowledge of the necessity of the laws of nature provisional. The point is that we should not worry what Hume pretended to worry, that is, that the actual laws of nature might change,

[14] In numerous works since his path-breaking paper (Friedman 1986), Michael Friedman has demonstrated how Kant constructed the Newtonian model of the motions of the solar system from the System of Principles of the first *Critique* plus some limited empirical observations about planetary motions as introduced in the *Metaphysical Foundations of Natural Science*. In discussion of this paper, he evinced concern that the present emphasis on the contingency of any conception of a system as a whole that we might have threatens this reconstruction. But in his original paper, he himself wrote that "not only is our constructive procedure strictly incompletable in principle for Kant, but even the limited success achieved so far – the construction of an *approximately* inertial system – depends entirely on fortunate, and contingent, facts about our initial a posteriori data ... Thus, unlike *mathematical* principles, which are 'constitutive with respect to *intuition*' (I can, e.g., instantiate the category of Quantity a priori in pure intuition by means of geometrical construction), *dynamical* principles of pure understanding are 'merely regulative principles of *intuition*' (B692)" (56–57). Thus, he himself has stressed an ineliminable element of contingency in his interpretation of Kant's reconstruction of Newton's system. And in many subsequent works, from Friedman 2001 to Friedman 2008, he has promulgated his conception of the relative a priori as those rules that are constitutive of objects *within* a particular scientific theory; this leaves open that the theory as a whole, let alone the always incomplete system of scientific knowledge of which it is supposed to be a part, may be contingent. Thus I believe I am expounding only Kant's own epistemological limitations on the kind of system-construction that Friedman has described.

but only that our claims to *knowledge* of the laws of nature might. Of course, the most general constraints on our knowledge represented by the pure forms of intuition and the pure categories of understanding are unrevisable, but since they are not sufficient conditions for our knowledge of particular laws of nature, they cannot guarantee that our claims to knowledge of the latter are unrevisable. From Kant's point of view, nothing can guarantee that.

There are also more particular limits to our knowledge of particular laws of nature built into Kant's conception of a system of such laws. It is clear that the descent from species to subspecies, whether we are thinking of subspecies of kinds or of laws, can never be complete, because no matter how finely a general concept is further divided, or in Kant's language determined, it can never approach the complete determination of an individual. Our general concepts can never fully determine individuals, as Kant argues §§76–77 of the "Critique of the Teleological Power of Judgment," sections that refer back to the general argument of the Introduction to the third *Critique* rather than to the specific theory of organisms into which they are inserted. As Kant says there, "Our understanding is a faculty of concepts, i.e., a discursive understanding, for which it must of course be contingent what and how different might be the particular that can be given to it in nature and brought under its concepts" (*CJ*, §77, 5:406). Because our concepts of natural kinds and natural laws, no matter how fine-grained, are always general, they always leave something about the particulars that fall under them contingent, and no matter how much further they are refined, this will remain true. So there can be no completeness in the descent to specificity in natural laws.

It would not seem as if the same must be true in the ascent to homogeneity or in the search for ever simpler concepts higher up in the system of concepts. For here nothing seems to block the ascent to a single kind or explanation as the basis of all of nature. But here there is a puzzle, for it might seem that if we were to succeed in ascending to a single concept and a single explanation for all of nature, it would become vacuous, and not explain any differentiation at all; indeed, on Kant's own model of the most fundamental explanation of natural phenomena, it seems as if we need at least two forces, attractive and repulsive forces, to explain differentiation in nature, in this case why we have neither an infinite dispersal of all bodies nor a concentration of all bodies in a single vanishingly small point but a distribution of bodies with determinate size, mass, etc. Moreover, Kant's illustration of fundamental explanatory factors in the Appendix to the first *Critique* – "pure earth, pure water, pure air, etc." (A646/B674), which smacks more of Milesian cosmology than of Kant's own – also suggests that we cannot ascend all the way to a single explanatory force or material but will need several in order to explain the diversity of nature, and indeed maybe not just two but three (earth, water, air) or even some indeterminately greater number that cannot be known a priori ("etc."). Perhaps this possible plurality of forms of matter does not threaten Kant's confidence that two and only two kinds of *forces* are necessary and sufficient to explain all of nature. Nevertheless, Kant's insistence that homogeneity is always only a regulative ideal ought to leave room for doubt whether we really have reached the most fundamental laws of the system. Even apart from the tension between his insistence that we must strive for a single ultimate law of nature and his insistence in his physics that we need two fundamental forces, he would seem committed to leaving it open whether we really know that *those* two are the most fundamental terms of explanation, thus that they might not be replaced by something else.

Thus incompleteness in specificity and uncertainty about homogeneity seem built into Kant's conception of a system of natural laws. Again, these do not seem to be academic concerns, to be left in the study when we go out for our evening of dinner and games, but to be ineliminable constraints on our knowledge of the laws of nature. It is in this way that Kant does not refute Hume's academic worry about induction, but rather replaces it with a substantive theory of the imperfection of our knowledge of natural law.

PART II

The Systematicity of Nature

CHAPTER 4

Why Must We Presuppose the Systematicity of Nature?

Hannah Ginsborg

4.1 Guyer's Challenge to the Presupposition of Systematicity

In the *Critique of Judgment* Kant argues that, in addition to the a priori principles of understanding presented in the first *Critique*, there is a further a priori principle associated with the faculty of judgment, that is, with the faculty for "thinking the particular as contained under the universal" (*CJ* 5:179). Unlike the principles of understanding, this is not a principle that we can know to obtain, but rather one that we must merely presuppose or assume. Relatedly, it is subjective rather than objective (e.g., *FI* 20:209; *FI* 20:21; *CJ* 5:184) and regulative rather than constitutive (*CJ* 5:180; *CJ* 5:194). However, like the principles of understanding, it qualifies as transcendental (e.g., *FI* 20:209; *CJ* 5:181 and passim). This principle is formulated by Kant in a variety of ways. Some passages characterize the principle as one of nature's purposiveness for, or conformity to, our cognitive faculties or our faculty of judgment. Others characterize it, with seemingly greater specificity, in terms of nature's systematicity, indicating that it can be expressed in such formulae as "Nature takes the shortest way . . . it makes no leap in the diversity of its forms . . . it is rich in species and yet parsimonious in genera" (*FI* 20:210; cf. *CJ* 5:182). But, however the principle is characterized, Kant makes clear that it is required for the exercise of judgment, specifically in its reflective rather than its determining capacity, that is, as finding universals for given particulars (*CJ* 5:179).

There are many problems regarding both the interpretation of this aspect of Kant's view and its defense. One interpretive crux concerns the exercise of reflective judgment for which the principle is required. On the less radical of two alternatives typically distinguished in the literature, the principle is required for pursuing systematic scientific inquiry, in particular constructing taxonomies of natural kinds and hierarchical systems of empirical laws. On the more radical alternative, it encompasses not just the systematization of empirical concepts and laws, but, more fundamentally, the kind of generalization required for arriving at empirical concepts and laws in the first place. Either way, we face a fundamental question: why does the exercise of judgment require us to assume a priori that nature is systematic? It would be one thing if the a priori principle of judgment merely enjoined us to seek systematic theories of nature. But the principle is not about our own cognitive aims but about nature itself. That feature of the principle opens Kant to a challenge, which has been articulated with particular force by Paul Guyer (1997: 42), namely, that the principle "does nothing but transform our own need for systematicity into a self-serving delusion." The most obvious justification for the principle as Guyer understands it is that, without it, our search for systematicity is not rationally motivated. But this justification, Guyer points out, is inadequate: the rationality of our search requires not a promise of success, but only the absence of a reason to believe that we will fail. Adopting the principle, then, seems to amount to "merely postulating or presupposing that an object will meet one's needs, rather than obtaining evidence that it does" (1997: 44).

Guyer formulates this difficulty in terms of the less radical interpretation of reflective judgment, but it also arises if, taking the more radical option, we think of reflective judgment as aiming at empirical conceptualization *überhaupt*. In the text most supportive of this option, Section V of the First Introduction, Kant formulates the principle as saying that "for all things in nature empirically determinate concepts can be found" (*FI* 20:211), adding in a footnote that, while this might seem to be a matter of logic, logic does not tell us that "for each object nature has many others to put forward as objects of comparison, which have much in common with it as to their form" (*FI* 20:211n). Here again, we can raise a version of Guyer's difficulty: why should the task of empirical conceptualization require us to presuppose in advance that nature is such as to make that task successful and, in particular, that things in nature display enough similarities that we can sort them into kinds? It is true that if nature did in fact display the kind and degree of heterogeneity that made it impossible for us to recognize things as having common features, presenting us with what Kant elsewhere describes as a "crude chaotic aggregate" (*FI* 20:209), we would be unable to bring objects under empirical concepts. But why does the activity of empirical conceptualization require us to rule out that possibility – a possibility that, at least on the evidence we have so far, does not obtain a priori? Why isn't nature's conceptualizability something that we come to discover through our success in conceptualizing it, rather than something that we have to assume in advance of our conceptualizing activity?

The more radical option gives rise to a further difficulty, namely, that it is not clear why empirical conceptualization should require the assumption of nature's systematic unity. According to Kant, "judgment, *which seeks concepts for empirical representations as such* ... must ... assume *for this purpose* that nature in its boundless multiplicity has hit upon such a division into genera and species as makes it possible for our judgment ... to arrive at empirical concepts, and their interconnection with one another, through ascending to more general but still empirical concepts" (*FI* 20:211n; emphasis added). But why is this assumption necessary? If we have to assume anything at all about nature in order to form concepts like *dog*, *tree*, and *water*, why can't it just be that there are natural kinds corresponding to these concepts, without any implication that these kinds can be further subdivided or in turn seen as subdivisions of higher-level kinds?[1]

In answer to the second difficulty, it might be claimed with Henry Allison that "the necessity for a hierarchical ordering in terms of genera and species follows from the very nature of a concept" (2001: 35). The concept *gold*, for example, is composed of further concepts, such as *yellow*, *metal*, and *soluble in aqua regia*; these concepts stand to *gold* as genus to species, but *gold* itself functions as a genus with respect to lower-level concepts of different kinds of gold, various things made of gold, and so on (2001: 35). Ido Geiger (2003) offers a more thoroughgoing justification of the need for systematic unity among concepts, drawing on the Sellarsian idea that since mere intuition cannot give content to our concepts, they can derive their content only from the systematic relations that hold among them. But even if it is right that we cannot conceptualize nature without thinking of our concepts as standing in systematic relations to one another,[2] it still does not appear to follow that conceptualization requires the presupposition that *nature itself* is systematically organized. So Guyer's initial difficulty remains.

Another proposal made by Allison suggests at least a partial response to Guyer's difficulty. This is that systematicity for Kant amounts to Hume's "uniformity of nature," so that, very roughly, the principle of nature's systematicity comes down to the principle of induction (2001: 38). This proposal is supported by a marginal note in which Kant says that Linnaeus could not "have hoped to outline a system of nature if he had had to worry

[1] Another possibility, on which conceptualization is restricted to shapes, colors, and tones, is considered by Guyer 1990: 29.

[2] For a challenge to this view, see Guyer 2003b: 294, n. 8.

that if he found a stone that he called granite, this might differ in its internal constitution from every other stone which nevertheless looked just like it" (*FI* 20:216n.). If Hume's problem of induction is the issue, then Guyer is not in a position to claim that we should look for evidence of nature's systematicity rather than assuming it a priori, since any attempt to look for empirical support for induction would be question-begging. However, Allison's proposal is called into doubt by Kant's claims that even though the principle of judgment cannot be "learned from observation," it can be "confirmed by observation" (*CJ* 5:186) and that the systematicity of nature is something in which we can feel pleasure (*CJ* 5:187) and a fortiori of which we can be aware. This suggests either that systematicity is not equivalent to the "uniformity of nature" in Hume's sense or that we can be aware of, and find empirical confirmation for, the uniformity of nature. Either way, Guyer's challenge remains: if the systematicity of nature is something that can be (even partially) confirmed in the course of empirical inquiry, what entitles us to assume it in advance of inquiry?

Guyer himself finds an answer to the challenge in two passages, from Sections IV and V of the published Introduction, where Kant points out that, even though particular empirical laws cannot be derived from the a priori principles of the understanding, they still must be regarded by us as necessary. What allows us to regard them as necessary, he says, is that we consider them "according to such a unity as they would have if an understanding . . . had given them for the sake of our faculty of cognition in order to make possible a system of experience according to particular laws of nature" (*CJ* 5:180). In other words, it seems to be the principle of nature's systematicity that allows us to regard particular empirical laws as necessary. According to Guyer, Kant's thought is that "an individual empirical generalization, which will seem contingent when considered in isolation, will appear to be necessary when it is embedded in a system of such generalizations" (2003: 287–288), especially if the system is one in which "generalizations at any level will appear to be entailed by the more general laws above them and confirmed by the more detailed laws beneath them" (ibid.). As Guyer notes, this is still not enough to establish that the principle is "a proposition about nature itself rather than merely an ideal for our concepts of nature" (ibid.), so still not enough to address the basic problem. But he argues that the problem can be addressed if we add a further assumption, namely, that we will often need to regard a generalization as necessary without possessing the whole system of generalizations that would entail and confirm it. Regarding such a generalization as necessary requires us to regard it as necessitated "by a system of regularities beyond our concepts, that is, by a system of regularities existing in nature itself" (ibid.).

In an earlier presentation of the same line of thought, however, Guyer presents the view in a somewhat more qualified way. He characterizes Kant as suggesting that recognition of entailment among the putative laws would "to some degree satisfy our demand for necessity even though we might still be able to imagine that the system as a whole could be replaced by some other, though equally systematic set of empirical laws" (1990: 41). The qualifications mark what seems to be a genuine difficulty: why should our conceiving of a law as embedded in some system of laws allow us to think of it as necessary if we do not in turn think of the whole system as itself necessary? We might regard a putative law's being embedded in a hierarchical system as sufficient for its lawlikeness if, following Michael Friedman (1992b), we also conceive of the system as anchored at the top by the a priori principles of understanding. But according to the line of thought Guyer is considering, it is the recognition of systematicity all on its own that is supposed to allow us to ascribe necessity to the putative laws.

One way to defend this line of thought would be to adopt a proposal defended by Philip Kitcher, on which systematicity not merely allows us to regard laws as necessary but is actually constitutive of their necessity. Kitcher, following Buchdahl (1969), sees Kant's endorsement of the a priori principle of systematicity as an answer to the problem of how we are justified in treating any regularity as necessary in the sense of lawlike – that is, counterfactual-supporting – as opposed to merely accidental. According to Kitcher, "Kant's solution to the puzzle of how we manage to recognize the necessity of laws is that . . . this necessity accrues to lawlike statements in virtue of their incorporation in a system that is constructed by following certain rules" (1986: 209), for example, rules enjoining maximum economy in higher-level explanatory principles and maximum diversity in the phenomena they explain. For Kitcher, this is not because incorporation in a system is an external indicator of necessity or a surrogate for it, but rather because it is simply constitutive of necessity. There is nothing more to being a law of nature or a natural kind than figuring appropriately in the ideal systematization of natural phenomena: laws just are "statements that play a particular role in the system that would emerge from an ideally extended enquiry" (1986: 215).

Kitcher's proposal, however, fails to do justice to Kant's repeated claims that the systematicity of nature is, at least as far as we can tell, contingent or, in other words, that we can conceive of the possibility that the laws of nature are not systematically organized. It is, Kant says, "surely possible in itself (at least as far as the understanding can determine *a priori*)" that the "manifoldness and heterogeneity of these laws, likewise the corresponding natural forms" could be "infinitely great," so that the laws "manifest a crude chaotic aggregate without the slightest trace of a system" (*FI* 20:209). Relatedly, the "harmony of nature with our cognitive capacity" is "recognized by the understanding as contingent" (*CJ* 5:185).[3] Kant's insistence on the contingency of nature's systematicity is of a piece with his denial that the principle of systematicity is constitutive, that is, that it legislates to nature: "reflective judgment can only give such a transcendental principle as a law to itself . . . and cannot prescribe it to nature" (*CJ* 5:180). If, as on Kitcher's view, the laws of nature were necessarily systematic, then the principle of systematicity would in effect be prescribed to nature: something that we could know a priori and not just something that we merely have to presuppose in order for reflection to be possible.

4.2 Nature's Purposiveness for Our Cognitive Faculties

I turn now to my own response to Guyer's challenge, which differs from the positions just surveyed by focusing not on Kant's formulations of the principle of judgment in terms of systematicity, but rather on what I regard as the more basic formulation in terms of nature's purposiveness for judgment.[4] I will offer an account of the presupposition of nature's purposiveness that makes plausible that it is a condition of the exercise of judgment, viewed as a capacity for empirical conceptualization. I will then argue that, given a plausible assumption about the nature of human

[3] Kitcher 1994 responds to an earlier version of this criticism (in Ginsborg 1992) by acknowledging the need to do justice to the contingency of nature's purposiveness for our cognitive faculties but continuing to maintain the constitutive connection between lawlikeness and incorporation in an ideal system. What is contingent is not that laws of nature constitute a system but that our current ways of theorizing approximate to that system. In effect, Kitcher distinguishes between the claim that nature's laws are systematically organized *tout court*, which he takes to be necessary, and the more demanding claim that their systematic organization is such that our present ways of theorizing approximate to a grasp of it, which he takes to be contingent. However, the passage just cited from *FI* 20:209 makes clear that we can conceive of natural laws as entirely lacking in systematic organization. More generally, Kant seems to regard formulations of the principle in terms of nature's systematicity as spelling out or making more explicit what is involved in nature's purposiveness for our cognitive faculties, so that it would be odd to suppose that he would combine an insistence on the contingency of nature's purposiveness for our cognitive faculties with a view on which systematicity was constitutive of lawlikeness.

[4] Other approaches that focus on purposiveness rather than systematicity are Floyd 1998 and Zuckert 2007.

concept-formation, the presupposition of nature's purposiveness for judgment commits us to the presupposition that nature is systematic.

Now it might seem that it should not make much difference whether we focus on purposiveness or on systematicity. Either way, it might seem, to adopt the principle of judgment is to presuppose something factual about nature, some objective feature in virtue of which it meets our cognitive needs. And, either way, Guyer's problem arises: what justification do we have for presupposing a priori something whose obtaining is discoverable – at least to some extent – by empirical inquiry? But I will argue that, even though the presupposition is about nature itself, it is not factual but normative. Moreover, it contains an irreducible reference to our own judging activity: it is about nature's relation to the very activity of reflective judgment for which the presupposition is required.[5] What we have to presuppose is a certain kind of normative fit between nature and our own judging of it, involving not only the idea that our judging is appropriate, or as it ought to be, with respect to nature, but also, and more fundamentally, the idea of nature's being essentially such as to call for that judging, or to make that judging appropriate. Briefly put, we must presuppose that nature *ought to be*, *should be*, or *is meant to be* judged by us in the ways in which we do in fact judge it.

In making this suggestion, I am extending, to the case of nature's purposiveness for our cognitive faculties, an interpretation I have offered elsewhere of Kant's notion of purposiveness in the biological and the aesthetic contexts.[6] According to this interpretation, purposiveness can be identified, very roughly, with normativity as such. More precisely, it can be identified with a notion of normativity considered in abstraction from any considerations of accordance with reasons or rationality. This is the very minimal notion of normativity captured by expressions like "ought" and "should" in contexts where they cannot be replaced with expressions like "has conclusive reason" to or explicated more indirectly in terms of an agent's reasons.[7] One motivation for this interpretation is to help us make sense of Kant's paradoxical-seeming claim that organisms are "natural purposes." This claim is problematic because the notion of a purpose seems to be associated for Kant with that of conscious design and to apply paradigmatically to artifacts, whereas Kant insists on the status of

[5] This is also emphasized by Floyd, e.g., in her characterization of the principle as one that "judgment gives itself *heautonomously*, i.e. circularly and self-reflexively" (1998: 207).

[6] See part III of Ginsborg 2015, especially essay 10.

[7] That there really is such a notion of normativity is up for debate. For defense, see Ginsborg 1998 and essays 10 and 15 of Ginsborg 2015.

organisms as products of nature. There seems to be a tension, then, between the thought of an organism as a purpose – canonically, "the object of a concept, in so far as the latter is thought as the cause of the former" (*CJ* 5:220) – and the thought of its being not a supernaturally produced artifact but something that comes to be through genuinely natural causes.

The solution to the difficulty about the idea of a natural purpose lies in attending to a certain feature of production by design, namely, that it has a normative aspect. The potter who sets out to make a vase has in mind a certain concept of what she intends to produce – the concept of an object made of certain materials, and with a certain size, shape, and color – and that concept plays a role in her production of the vase. But the role it plays is not just causal but normative: the concept figures in her mind as a normative constraint on how the vase is to turn out, as determining how it *should be* or *ought to be* or *is meant to be*. If it turns out in such a way that the concept does not apply to it – if, say, it comes out of the kiln with a green color rather than the yellow she intended – she will think not merely that it is green rather than yellow, but that, in being green, it falls short of what she had in mind, that it is not the way the vase is meant to be or supposed to be. The normativity implicit in her attitude toward the vase is independent of reasons. She need not have had any reason to choose yellow rather than green for the color of the vase, and she may actually prefer the green color to the color she intended. But that does not stop her, or us, from thinking of the color in normative terms, as "failing to conform" to the standard she had in mind and as representing a "mistake" or an "error" in the production process.

Appeal to this "thinly" normative aspect of design – "thin" because it makes no reference to reasons – allows us to understand how natural objects, for Kant, can be thought of as purposive without undermining our conception of them as products of nature rather than design. To think of an object as purposive, we do not have to think of it as produced by an intelligent being so as to conform to a normative constraint: we can leave the intelligent being out of the picture, and simply think of the object as conforming to a normative constraint. Although, then, we get our initial handle on the notion of purposiveness by reflecting on what is involved in actual production by design, the core of the notion is not the idea of design itself but rather that of the normativity implicit in design. Kant suggests this in a passage where he characterizes teleological judgments about natural things in explicitly normative terms, saying that such a judgment "compares the concept of a product of nature as it is with one of what it *ought to be*" (*FI* 20:240). He points out that we recognize, of an eye, that it is something by means of which we can see, and of a stone, that it is suitable for building. But, he says, "it is only of the eye that I judge that it *ought* to [*sollen*] be suitable for seeing" (ibid.). Like an artifact, an organic being is something of which we can say not merely that it is a certain way, but that it ought to or is meant to be that way.

The purposiveness just described is what Kant calls objective purposiveness, which is a matter of how an object, artificial or natural, ought to or is meant to *be* (or, relatedly, what it is meant to *do*). But the purposiveness of nature for our cognitive faculties is not objective, but subjective: a matter not of how nature ought to or is meant to *be tout court* but of how nature ought to or is meant to *be judged*. By way of transition to this latter idea, we might return to the example of artifacts and note that we can speak not only of how a vase ought to or is meant to *be* – hollow, tall enough to accommodate the stalks of flowers, stable, impermeable to water – but of how it ought or is meant to *be used*. A vase is not just something that facilitates the preserving and displaying of cut flowers but something to which that use is, in a sense, appropriate or fitting. Someone who stores pennies in the vase or uses it as a water jug is, in the same sense, not using it as it ought to be used, or not making proper use of it. The sense of normativity here, as in the case of how the vase is meant to be, is independent of reasons. There is no reason to refrain from using a vase as a penny-jar, a jug, or – if the vase is attractive – as an ornament. But the flower-displaying use remains privileged in a sense that is at least possible to describe using normative terms like "should" and "ought." Even though in most contexts talk of how a vase should be used or, more generally, treated, has a prudential or moral content – that vase shouldn't be taken on the camping trip because it's valuable or because we promised the owner we'd be careful with it – we can still speak of certain uses as corresponding or failing to correspond to how the vase "should" be used in the thin sense that is most naturally captured by idioms like "meant to" and "supposed to." This is especially clear where the relevant behavior is more complex. We speak of how a violin should be or is meant to be played – roughly by drawing the bow across the strings between the end of the fingerboard and the bridge – without any implication that one has reason not to make sounds on it in other ways. For example, a composer might specify that the player hit the bow against the strings or play below the bridge, and in such cases part of what makes the result aesthetically interesting is that the violin is not being played in the way in which – in the thin sense – it should be played.[8]

[8] An extreme case is Helmut Lachenmann's 1969 *Pression* for solo cello, in which – quite deliberately – the cello is never played as a cello "should be" played.

I have been using the example of artifacts to illustrate the general idea of a normative fit between an object and certain behavior where the object not only facilitates the behavior but also makes it appropriate. While Kant does not explicitly discuss this kind of normative fit between a particular object and our overt behavior in relation to it,[9] I think he does invoke an idea of the same kind of normative fit between an object and our psychological response to it: namely, in the idea of the subjective purposiveness that we ascribe to an object when we make a judgment of beauty.[10] Such a judgment, on my reading, consists in the subject's responding imaginatively to an object in a way that involves her taking that very response to be appropriate to the object, but without the thought that she is according with a rule or concept that determines that response as appropriate. In contrast to the case of an objective judgment about the object, where she might take herself to be responding appropriately in virtue of the fact both that the object has a certain objective property and that she is judging it to have that property, she simply sees the object as normatively calling for that response, or as making it appropriate. The property she ascribes to it in having that response – that of making appropriate the very response she is having – is ineliminably normative and subjective. It is not that she takes there to be some objective fact about the object that gives her a reason to respond to it in the way she does or makes it the case that her response is veridical; rather, in responding to the object, she perceives it in a way that cannot be further analyzed, as simply meant to be responded to in that way. This is what it is, as I understand Kant, for her to perceive the object as subjectively purposive in the sense relevant to judgments of beauty.

There are important disanalogies between the aesthetic case and that, say, of the violin. For one thing, the activity of playing a violin does not essentially involve taking what one is doing to be appropriate to the violin; whereas as I understand the judgment of beauty, one's imaginative response to the object essentially involves awareness of the appropriateness of one's response. For another, we can specify how a violin ought to be played, that is, give what might be regarded as rules for playing a violin, or standards for violin-playing, whereas this kind of specification is not possible in the case of one's imaginative response to the beautiful object, where one can specify how one ought to respond to a given object only by saying that it is *this way*. What the violin example is intended to bring out

[9] Although it might be seen as a form of "relative" or "extrinsic" objective purposiveness (see *CJ* §63 and §67).

[10] See Ginsborg 2015: 90 and 248.

is simply the idea of a certain kind of normative fit between a thing and one's response to it, where the normativity is of a "thin" kind that does not involve the thought that one has reason to respond in the way one does. That idea is central to my understanding of nature's purposiveness for our cognitive faculties, to which I now return.

4.3 The Principle of Nature's Purposiveness as a Condition of Empirical Conceptualization

I suggested above that the a priori presupposition of nature's purposiveness for judgment amounts to the presupposition that nature ought to be, should be, or is meant to be judged by us in the ways in which we do in fact judge it. I went on to try to clarify the notion of normativity invoked in that formula and to indicate how it can be understood as figuring in Kant's account of the purposiveness of artifacts, organisms, and especially objects of aesthetic judgment. I now want to explain why the presupposition, so understood, is required for the exercise of judgment.

Following Kant's very general characterization of judgment as "the capacity to think the particular as contained under the universal" (*CJ* 5:179), I adopt the more radical of the two options distinguished in Section 4.1, taking reflective judgment to aim not just at the systematization of concepts and laws, but at empirical conceptualization *überhaupt*. Our need for the principle of purposiveness stems from a claim to normativity that I take to be built into the very activity by which we come to arrive at empirical concepts. Roughly, on my view, we bring objects under empirical concepts by exercising the same kind of sorting or discriminative capacities possessed by animals. What makes our exercise of those capacities different from those of animals – what makes it genuine conceptualization – is that, in sorting the objects as we do, we take what we are doing to be appropriate to the objects. We don't just sort them together in a particular way – say, sorting the dogs together or the green things together – we take it that this is how we should sort these objects. I take this to be part of what Kant has in mind in the first *Critique* when he describes concepts as rules for synthesis, in particular the synthesis of reproduction in imagination described in the first edition Transcendental Deduction.[11] Our reproduction of past representations to form the perceptual image of a dog involves calling to mind elements from our representations of previously perceived dogs, and so amounts to a kind of sorting of

[11] See essay 3 of Ginsborg 2015.

the presently perceived dog with other dogs. That the features that come to mind are of previously perceived dogs, as opposed to previously perceived cows, is a result not of our already having grasped a concept that guides our reproduction of past representations, but rather of natural tendencies to associate representations in certain patterns rather than others, tendencies that are by and large shared with animals. Unlike animals, however, we are conscious, in calling to mind past representations, of what we are doing as appropriate to our present circumstances. That consciousness distinguishes our synthesis of representations from the "blind" association carried out by animals. It makes it the case that we regard it as governed by a rule, which is – on my understanding of Kant – just what it is to think the individual we are perceiving under a concept.

According to this line of thought, it is implicit in our activity of empirical conceptualization – or, in the terms of the third *Critique*, reflective judgment – that it involves the consciousness of that very activity as appropriate to the objects with respect to which it is carried out. Without that consciousness, we would not be conceptualizing at all. Rather, our reflection would be of the kind that Kant describes as "going on . . . in animals, although only instinctively, namely not in relation to a concept which is thereby to be attained but rather in relation to an inclination which is thereby to be determined" (*FI* 20:211). Here it is important to note that the relevant sense of appropriateness cannot be spelled out in terms either of veridicality or of rational justification. It is not that we recognize that Fido is a dog and hence that, in conceptualizing him as a dog, we are making a judgment about Fido that is true rather than false. Nor is it that we recognize that Fido is furry and barks, and so should be sorted with the other furry barking things as opposed to the mooing things. While we can justify the claim that Fido is a dog, viewed as an exercise of determining judgment, by saying that Fido barks and is furry, and that observation has shown us that furry barkers are dogs, there is no comparable justification for the exercise of reflective judgment that yields the concept *dog* rather than *dog-or-cow* or *green* rather than *grue*. As Kant puts it, reflective judgment has to proceed "not schematically, but technically, not as it were merely mechanically . . . but rather artistically" (*FI* 20:213–214). Empirical conceptualization, viewed as making concepts possible rather than as the application of concepts we already possess, has to proceed without the benefit of antecedently grasped rules that tell us how it ought to be done, although, as in the case of artistic production, this does not preclude a consciousness of ourselves as normatively constrained in our activity. Even if the artisan does not already have in mind detailed specifications for something she is making, and at some point comes to a point in her work where she has no rule to tell her how to proceed, she can still think of what she does as the right or appropriate thing to do in that context, or as according with a standard that cannot be specified until after the work is completed (and perhaps not even then). Similarly, in empirical conceptualization, we think of our sorting as governed by a standard determining how the objects should be sorted, even though we cannot specify that standard except in terms of the concepts yielded by the activity of sorting itself.

These considerations suggest a parallel between the reflection on nature that yields empirical concepts and the reflection on a particular object in virtue of which we judge it to beautiful. In both cases there is an imaginative activity that involves the consciousness of its own appropriateness with respect to the object (or objects) toward which it is exercised. And in both cases that consciousness does not depend on the prior consciousness of rules specifying how we ought to carry it out. In the aesthetic case I used this idea to explain the subjective purposiveness involved in making a judgment of beauty. To perceive the object as subjectively purposive is to perceive it as normatively calling for, or making appropriate, our response to the object. I am drawing on the same idea to explain nature's purposiveness for judgment. To take nature to be purposive for our exercise of reflective judgment in conceptualizing nature is to take it to make appropriate our ways of conceptualizing nature. Our justification for regarding nature in this way is that, if we did not, we would not be able to regard our conceptualizing as appropriate with respect to nature, which would mean that it would not amount to conceptualizing at all. There would be no need for the presupposition of purposiveness if, as in the case of determining judgment, we could think of what we were doing as appropriate in the sense of veridical or justified. That would be a way of thinking of our activity in normative terms, as fitting nature, without having to ascribe anything normative to nature itself. We can think of the judgment that Fido is a dog as appropriate to nature without thinking of nature as meant to be judged by us in such a way that we judge Fido to be a dog. It is enough to think of nature as being such that Fido is a dog, or, in short, to think that Fido is a dog. But in the case of reflective judgment there is no alternative to thinking of the appropriateness involved in our own cognitive activity with respect to nature in terms of a corresponding normative feature of nature with respect to our activity. To think of ourselves as judging appropriately we have to think of nature as making our judging appropriate, in a sense that involves the ascription of something irreducibly normative not only to our own judging but to nature as well.

4.4 From Purposiveness to Systematicity

I want now to address two lacunae in the account offered so far. First, in saying that we have to presuppose that nature is meant to be conceptualized by us in the ways in which we in fact do conceptualize it, I seem to have left no room for the obvious fact that we frequently revise our concepts. Second, I have said nothing about Kant's formulation of the principle in terms of the systematicity of nature, although it is that formulation that most clearly generates the difficulty described in Section 4.1. To address both of these lacunae, I want to be clearer about the claim that we must think of nature as meant to be conceptualized in the ways we conceptualize it. That claim, as I understand it, does not rule out the thought that we might come to reject particular concepts and systems of concepts in favor of concepts that we find more appropriate to nature than the ones we had before. What it does rule out is two possible thoughts, neither of which can be entertained by us without undermining the claim to appropriateness built into our conceptualizing activity. The first is that there is no such thing as an appropriate way of conceptualizing nature: that any one way of sorting the natural things presented to us is as good as any other. The second is that, while there are appropriate ways of conceptualizing nature, these are completely at odds with our natural ways of conceptualizing. Neither thought can be rejected on empirical grounds. In particular, the second is quite consistent with the idea that our natural dispositions to sort things have so far enabled us, by and large, to make predictions that turned out to be successful, and indeed that our ways of sorting will continue to facilitate successful predictions indefinitely. But if we allow either thought as a possibility, then we cannot keep hold of the idea that, in sorting objects as we are naturally inclined to sort them or, so to speak, as nature invites us to sort them, we are sorting them in a way that is also appropriate to nature. To borrow a phrase from Kant, we have no alternative but to think of our sorting as "random and blind, and without legitimate expectation of its agreement with nature" (*FI* 20:212).

So understood, however, the principle is not so strong as to rule out the possibility of our revising our present system of concepts. It requires not that our natural sorting inclinations correspond immediately to the ways in which natural objects would be sorted by an ideal human enquirer, only that they tend to take us in the direction of those ways. And this is something that we in fact take for granted in the course of empirical scientific inquiry. For our revisions of the concepts initially yielded by our natural ways of sorting things are themselves arrived at through procedures that make use of those basic ways of sorting and rely on the assumption of their appropriateness. When, for example, we come to classify whales as mammals instead of fish or to distinguish what we earlier thought of simply as jade into jadeite and nephrite, we are still following our natural sorting inclinations. The difference is just that these inclinations are having their effects in a different context. In broadening the concept *mammal* to include whales and other marine mammals, we are still relying on a natural sense of what is similar to what, but as applied to anatomy and mode of reproduction as opposed to shape and habitat. Similarly, in coming to distinguish different chemical substances that initially strike us as belonging together, we will often rely on experiments whose results we interpret in ways corresponding to our more basic sorting inclinations – for example, when we use the color of a piece of litmus paper to determine whether something is an acid or a base.

There is also a second respect in which, in revising our systems of concepts, we rely on our natural sorting inclinations. This is that we are naturally inclined to sort things not only at the "basic level" and in ways corresponding to simple features like color and shape, but also at various different levels of specificity. We do not – and this is something that again comes naturally to us – rest content with just sorting Fido together with the other dogs. We also sort him more specifically with the poodles and, if we are dog fanciers, with a particular variety of poodles. In the other direction we sort him with other animals and more generally with living things. It is of a piece with this feature of how we sort that we are inclined to prefer ways of sorting that are more conducive to systematic classification. This is part of what drives us to revise our initial classifications. Our carrying out such revisions with the aim of arriving at an increasingly systematic classification of nature is itself a natural feature of our sorting behavior, and it is part of what makes it the case that we do not stop at the conceptualizing that is required for ordinary experience but rather go on to engage in scientific inquiry.

This second respect points to a way of addressing the other lacuna, regarding the connection between purposiveness and systematicity. Suppose we assume, as suggested above, that it is a structural feature of our ways of conceptualizing nature – itself a way, albeit a higher-order way, of conceptualizing nature – that we conceptualize in ways that are conducive to systematic theories of nature. Then our presupposition that nature is meant to be conceptualized as we in fact conceptualize it will imply the presupposition that nature is meant to be conceptualized in a way that is systematic, which in turn amounts to the presupposition that it is systematically organized. Presupposing this is presupposing something about nature and not just about our cognitive faculties. Moreover, the content of the presupposition is factual rather than normative: it says something about how nature actually is, something that can be at least partially confirmed by our experience of nature. But it is not vulnerable to Guyer's challenge, because we can show how it follows – given a further assumption about the nature of our own conceptualizing activity – from a principle that is not itself factual and that, if the argument sketched in Section 4.3 is correct, is a genuine a priori condition of reflective judgment.

How does this solution to Guyer's challenge compare with Guyer's own defense of the principle of systematicity in terms of the need to regard empirical laws as necessary? While I have been focusing on reflective judgment as a capacity for recognizing objects as instances of empirical concepts rather than as a capacity for recognizing events as instances of empirical laws, the solution I have proposed is consistent with the idea that the principle is needed in order to recognize the necessity of empirical laws. This is because there is an intimate connection between the thought that our natural ways of sorting things are appropriate to nature and the thought that the regularities we perceive among phenomena, qua instances of the corresponding concepts, are lawlike. For example, we find it natural to sort things in ways corresponding to the concepts *salt* and *water* as opposed to the artificially constructed concepts *stuff in a blue cylindrical container in my kitchen* and *stuff that comes out of the tap*, and so we think of *salt* and *water* as corresponding to ways in which the phenomena we observed ought to be sorted. That attitude is of a piece with our taking the statement "salt dissolves in water" to express a lawlike generalization, in contrast to the merely accidental generalization expressed by "the stuff in blue cylindrical containers in my kitchen dissolves in the stuff that comes out of the tap." It is our tendency to regard different samples of salt and water respectively as meant to be grouped together – as members of the same natural kind – that disposes us to regard various observed instances of salt's dissolving in water as manifesting a lawlike regularity.

My account of the role of judgment in recognizing the necessity of empirical laws is, however, unlike Guyer's in appealing not to systematicity as such, but rather to the more basic notion of purposiveness. It is true that the regularities we regard as necessary are typically also regarded by us as systematically related to higher-level regularities. However, that is not because systematicity as such is a criterion of lawlikeness. Rather, it is because – as a matter of fact – we are disposed to conceptualize in a way that is conducive to a systematic understanding of nature, so that we end up regarding as lawlike just those regularities that we can also incorporate into an explanatory system. There is thus no need to follow Kitcher in taking systematicity to be constitutive of lawlikeness, with the consequent risk of undermining Kant's view that a regularity could be lawlike and still resist incorporation into a systematic theory.

One might still worry, though, that the difficulty I raised for Kitcher about systematicity recurs for the more basic notion of purposiveness. It is essential, if we are to preserve Kant's commitment to the regulative character of the principle of judgment, that we can make sense of the idea that nature's purposiveness for judgment is contingent. On my reading, that amounts to saying that we must be able to acknowledge the possibility of a mismatch between our ways of conceptualizing nature (and, correspondingly, the regularities we conceive as lawlike) and ways in which nature ought to be conceptualized (and, correspondingly, the regularities that genuinely are lawlike). Can we, while assuming that our ways of conceptualizing are appropriate to nature, also make room for the possibility that natural things ought to be conceptualized in ways that are quite different from those that come naturally to us? Or does my reading build into the concept of a natural kind or a natural law that it accords with our ways of conceptualizing, so that we cannot conceive of natural kinds or lawlike regularities that cannot be grasped, as such, by human beings?

The answer to this difficulty lies in recognizing the essentially first-personal character of the principle of purposiveness. The presupposition that nature ought to be conceptualized in the ways we conceptualize it makes reference to our conceptualizing activity not third-personally, qua the conceptualizing activity of human beings, but first-personally, qua *our* conceptualizing activity. We assume, in our activity of conceptualizing, that nature calls for, or makes appropriate, *this very activity*. But even though we cannot exercise reflective judgment without conceiving of a normative fit between our activity of reflective judgment and nature, we can still step back and adopt a third-personal view on the relation between that activity, now conceived as the actualization of human sorting dispositions, and the ways in which natural things ought to be sorted. And from that point of view, the concept of a way natural things ought to be sorted comes apart from the concept of how human beings in fact sort them. We get the idea that there so much as *are* ways in which nature ought to be conceptualized – that there are, in other words, genuine natural kinds and corresponding empirical laws – only through taking our own ways of sorting natural things to be appropriate to nature, and thus ruling out the possibility of a radical mismatch between our ways of conceptualizing and ways in which nature ought to be conceptualized. But once we have the notions, on the one hand, of a way in which nature ought to be conceptualized, and, on the other, of the way in which human beings are naturally inclined to conceptualize it, we can see how the two might come apart. It is possible, then, to see how the principle of purposiveness is "subjectively necessary" (*FI* 20:209) for the exercise of reflective judgment – something that, in our capacity as conceptualizers, we have to take for granted – while at the same time allowing that the principle, conceived as making a third-personal claim about the relation between human cognitive capacities and the natural environment to which they are directed, is "objectively contingent" (*CJ* 5:185).

Acknowledgments

Thanks to Daniel Warren, for helping me think through the ideas in this chapter, and to Angela Breitenbach and Michela Massimi, for their very insightful comments on earlier drafts. Thanks also to the other participants in the Edinburgh *Kant and Laws* conference, and to the participants at Kant conferences at Southampton and McGill where this chapter was subsequently presented, for helpful discussion.

CHAPTER 5

Empirical Scientific Investigation and the Ideas of Reason

Rachel Zuckert

In the Transcendental Dialectic of the *Critique of Pure Reason*, Kant argues that traditional special metaphysics is, necessarily, a failed enterprise; because of the nature of our cognitive capacities, we cannot in principle attain knowledge of the objects of the rational ideas – the soul, the world as a whole, or God. In the Appendix to the Transcendental Dialectic, Kant argues that these ideas nonetheless do have a positive epistemic role. They serve as a "focus imaginarius" or "schema" for the regulative principles that guide empirical scientific investigation toward attaining systematically unified science.[1]

It is worth pausing to recognize the strangeness of this claim. On Kant's view the ideas not only provide no knowledge, but also are illusory, leading us to try to attain knowledge that is in fact beyond us. Relatedly, they cannot in principle be instantiated in experience (see A327/B383), and, Kant claims, on closer investigation they are revealed to be nearly empty of content altogether. It is, then, unclear what such ideas could contribute to scientific investigation, which is of course concerned to find contentful truths about objects given in experience (not empty illusions about things beyond experience). It is unclear, specifically, how they could help us to apply the regulative principles to empirical particulars in investigation, as Kant suggests in using the term "schema." Or, more commonsensically: why should a priori conceptions of the soul, God, or the world as infinite whole have any role to play in empirical scientific investigation?

I attempt here to make some sense of this puzzling claim. I shall propose that it is precisely *as* representations of unknowable objects, and *as* nearly empty, that the ideas are useful for scientific investigation on a Kantian view. Specifically, I shall suggest that the ideas function as "optimistic placeholders": they function as encouragement to investigation, suggesting

[1] A644/B672, A674/B702. Though Kant uses this term differently, I use "science" to refer exclusively to empirical natural science.

(if illusorily) that *there is* something "out there" to be found in ongoing empirical investigation. Precisely because they are nearly empty, however, the ideas do not predetermine the results of that investigation; they are *mere* placeholders for empirical results of properly empirical investigation. Because they present objects that will never be known by empirical science, moreover, the ideas also mark the limits of empirical scientific investigation as such, indicating that it will never be fully adequate to the demands of reason. Thus, I shall suggest, the ideas serve as schemata for – aids for the application or (perhaps better) proper employment of – the regulative principles by (paradoxically) *preventing* their application directly to the empirically given.

5.1 A Reminder: Ideas, Dialectic, and the Regulative Principles of the Systematic Unity of Science

I begin by recalling the basic contours of Kant's account of the ideas. In the Dialectic, Kant is largely concerned to show reason's propensity to dialectical error, the way in which it leads us to lay claim to knowledge that is in fact forever beyond us. Such claims purport to concern the non-empirical objects of three transcendental ideas formed by reason: the soul, the world as a whole, and God. But such ideas, Kant claims to establish, do not truly present us with objects. Rather, they are mere, illusory hypostatizations (objectifications) of the rational demand to attain complete knowledge and ultimate explanations – or, in Kant's terms, to find as "given" the unconditioned condition(s) for any conditioned item.[2] In formulating the ideas, and in claiming acquaintance thereby with their transcendent objects, reason purports to have satisfied its own demand, to have found such given unconditioneds. Reason is then enticed to make claims a priori about them (that the soul is a simple substance, that God created the world and so forth), that is, to engage in special metaphysics. But, Kant argues famously, reason tries in vain. All three ideas, he claims, are nearly empty: they present us with *mere* aspirations, which are *merely* projected as objects. Correspondingly, we can attain no knowledge of their objects; all such purported knowledge rests, Kant argues, on mistaken assumptions and fallacy or entails contradictions.

In the Appendix to the Dialectic, however, Kant argues that despite its propensities to transcendent-metaphysical error, reason nonetheless has a positive epistemic role to play in our legitimate and (at least potentially)

[2] A308/B364. See Grier 2001: 117–139.

successful *empirical* cognitive enterprises. Reason formulates the conception of the systematic unity of science as the ultimate goal of empirical scientific investigation (and one that would satisfy rational demands for totality and unity). And reason sets forth regulative principles that guide investigation toward accomplishing that aim: the principles of homogeneity, specificity, and affinity enjoin us, respectively, to seek more universal empirical laws, more specific laws, and continuity among those laws. In other words, these principles instruct us to fill out a complete system of empirical laws, whereby we may understand all aspects of objects according to such laws, and thus attain "thorough-going determination" of particulars in their empirical characters (A656/B684). (In the following, I use "regulative principles" to refer to these three principles.) Unlike the constitutive principles formulated by the understanding (and defended in the Transcendental Analytic), these regulative principles cannot be *known* to be true of objects of experience. For all we can establish, on Kant's view, nature might not be fully systematic, and the complete systematicity of science seems in fact on Kant's view to be forever out of reach (see A663/B691). Nonetheless, only by following such principles can we *approximate* "asymptotically" to the best (most systematic) science (ibid.).[3]

If one acknowledges the positive epistemic role played by these regulative principles, Kant claims, one can in turn identify a positive epistemic role for the ideas of reason. Namely, they are to serve as schemata of these regulative principles,[4] not presenting objects that might themselves be known, but rather (somehow) facilitating the application of the regulative principles to experiential objects, for the sake of systematic science (A670/B698–A671/B699). In particular, the idea of the soul will enable us to "connect all appearances, actions, and receptivity of our mind to the guiding thread of inner experience *as if* the mind were a simple substance that (at least in this life) persists, while its states . . . are continuously changing" (A672/B700). The idea of the world prompts us to "pursue the conditions of the inner as well as the outer appearances of nature through an investigation that will nowhere be completed, *as if*

[3] This aspirational, not-guaranteed status of systematic knowledge of nature is subsequently emphasized in the *CJ*. Owing to space constraints, I will not discuss here the relation between the Appendix and *CJ* discussions.

[4] See A674/B702. Kant does not here use the plural "schemata," but does refer to all three ideas in writing that "their reality should hold only as that of a schema of the regulative principle for the systematic unity of all cognitions of nature," or again (later in ibid.) as furnishing a "schema" for the "systematic unity." He does, however, tend to use the locution "schema" for systematic unity more frequently for the idea of God specifically than for the other two ideas (see A697/B725, A699/B727).

nature were infinite in itself" (ibid.). And in taking nature to be organized systematically, we are interpreting it "as if" created by God, as if intentionally ordered by an intelligent and beneficent creator (A672/B700–A673/B702, A688/B716).

I shall return below to the apparent differences among these schematic-regulative functions for the ideas. For now, I note that in each case Kant emphasizes that we (still) know nothing about the purported transcendent objects of the ideas. Nor, he emphasizes, do we "derive" anything about empirical nature from them (A673/B701). Rather, we consider nature only "as if" it is related to those ideas or their objects (as emphasized in the passages just quoted). Kant correspondingly expatiates on the stultifying effects on empirical investigation that result from misusing the ideas, particularly the idea of God. If we lay claim to knowledge of God's connection to nature, we will fall prey to the epistemic diseases of "lazy" and "perverted" reason, that is (respectively), of neglecting to explain nature by recourse to natural scientific laws in favor of simply pointing to God's intentions, and of coming up with arbitrary explanations based on our purported knowledge of those intentions (A689/B717–A694/B722). In both cases, we exempt ourselves from the need to investigate nature empirically, claiming instead already to know. The ideas of reason are, in sum, to serve not as subject matters of knowledge in their own right, but as "heuristic ... concepts," useful as guides for empirical scientific investigation, specifically as guides (schemata) to using *other* guides (the regulative principles) that are to direct us toward achieving systematic empirical knowledge of nature (A671/B699).

5.2 The Question: Why (and How) the Ideas of Reason?

There has been considerable scholarly discussion concerning why and to what degree, for Kant, systematic unity *is* a necessary goal for scientific investigation – and thus in turn as to the status of the regulative principles. Kant often characterizes systematic unity as an ideal we should try to realize as far as possible. But he also makes stronger claims: systematic unity (or approximation to it) is the "sufficient mark of empirical truth" or necessary for us to "have" (employ) understanding at all.[5] Pursuing systematic unity of science (or adhering to the regulative principles) might, then, be a necessary condition for the possibility of experience or

[5] A651/B679 and A657/B685; see also A660/B688–A661/B689, A680/B708. See also *CJ* 5:184–185.

empirical knowledge on Kant's view – though it is not clear why or how this should be.[6]

I shall remain neutral concerning how to answer these questions here, however, so as to consider the yet more controversial part of this Kantian doctrine. That is: *allowing* that we aspire to systematic science, and that empirical investigation therefore ought to be guided by the regulative principles, do we really need the ideas of reason for that enterprise?[7] As noted, this is a striking claim: why ever, in investigating *empirical* matters precisely as such, in aiming to extend *empirical* science, should we employ ideas of that which lies beyond experience? (Perhaps because this claim is so puzzling, commentators interested in Kant's claims concerning systematicity tend not to attend much to it.)[8]

We may also ask, more specifically, why does one need schemata – further heuristics, further guidance – to follow the regulative principles? Why can one not simply aim to find more universal or more specific laws, while attending just to nature, to the empirically given?[9] More technically, the ideas seem not to fit Kant's understanding of schemata particularly well. Kant introduces the term "schema" in the Transcendental Analytic to refer to a "mediating representation" of the imagination, a "third thing . . . [between] the category on the one hand and the appearance on the other, [which] makes possible the application of the former to the latter" (A138/B177). The schemata are meant, generally speaking, to explain how (universal) concepts may be applied to (particular) intuitions, or how knowers recognize sensibly given instances of concepts.[10] In the case of the categories, the schemata also give those (abstract, nonempirically derived) concepts some sensibility-related content such that they may be applied determinately to objects given within sensible experience (A137/B176–A138/B177).

None of this appears suited to describe a relationship between the regulative principles and the ideas. On Kant's view, the regulative

[6] Kant appears both to deny and to assert that the regulative principles are principles of the possibility of experience or empirical cognition (e.g., A509/B537, A663/B692), or that the ideas may have a transcendental deduction (A633/B692, A671/B699). For recent discussion of these puzzling reversals, see McLaughlin 2014a.

[7] This division of questions corresponds to the two sections of the Appendix, the first of which treats the regulative principles of systematic unity, while the second concerns the (purportedly related) role of the ideas of reason in empirical scientific investigation.

[8] E.g., Kitcher 1986, Rohlf 2014, Rush 2000. Notable exceptions are Grier 2001 and Rauscher 2010.

[9] The Introductions to the *CJ* might be read to suggest so; there Kant returns to the project of empirical systematization, but (with one exception quoted below) hardly mentions the ideas.

[10] For representative opposing views concerning the significance of the schemata in Kant, see Bennett 1966 and Longuenesse 1998.

principles *are* a priori, like the categories: they are demands made by reason, standards for fully satisfactory knowledge, in accord with which we then investigate the empirically given.[11] But the ideas are not mediating representations that lie (somehow) between such a priori demands and the sensibly given: inasmuch as they purport to represent objects beyond experience, they would appear considerably *further* from the empirically given than are the injunctions of the regulative principles.

Nor do the ideas seem able to provide further content and in that sense aid in the employment of the regulative principles. These principles *are* vague; they give little direction about what to look for, what sorts of higher or lower laws there might be. Their indeterminacy is, indeed, one reason they are *regulative* principles, that is, *mere* guides to investigation, which do not articulate any characteristics an object might have (A661/B689).[12] One might well wonder how to do what these vague principles enjoin, and perhaps desire further guidance – how *am* I to seek a more universal or more specific law, where should I look? But again, the ideas of reason cannot address this concern: one place we can be sure that we should *not* look for explanations of the empirical, on Kant's view, is beyond experience, as he reminds his readers frequently in the Appendix.[13] The ideas of reason are also, as noted above, themselves nearly empty of content (as Kant also emphasizes in the Appendix),[14] and thus unlikely to furnish content for the regulative principles.

In sum: how can the ideas be necessary for empirical scientific investigation, on Kant's view? Does one, specifically, need schemata to aid in employing the regulative principles that are to guide such investigation, and if so, why are the ideas – representing objects beyond the empirical – plausible candidates?

As noted, I shall suggest that it is precisely the (near) emptiness and beyondness of the ideas that suits them to serve as schemata for the regulative principles. I begin, however, by returning to a subordinate interpretive point mentioned above, namely that – at least at A672/B700

[11] See A651/B679, A657/B685, A661/B689–A662/B690.

[12] The cited passage concerns the principle of affinity in particular, though I take it that this point applies to all three principles.

[13] Kant's resolution to the Fourth Antinomy might be taken to press this point: empirical investigation proceeds according to its own principles, "from empirical conditions to higher ones, which are likewise always empirical," and this is "not at all opposed to" the possibility of an entirely different sort of condition for the entire series (i.e., God) (A564/B592). Thus, one might think, the idea of God has nothing to do with, and so can play no guiding role with respect to finding, the members of the empirical regress. Kant's language at A673/B701–A674/B702 concerning the regulative use of the idea of God is (surprisingly) reminiscent of that discussion.

[14] See A674/B702; specifically, about the idea of the soul, see A683/B711–A684/B712; the idea of world, see A677/B705; and the idea of God, see A678/B706–A681/B709.

(as quoted above) – Kant suggests three rather different regulative roles for the ideas. I shall propose that these three functions together comprise the regulative function of the ideas as "focus imaginarius" – Kant's famous metaphor at A644/B672 – and this, I shall propose in turn, can help us to understand how the ideas function as schemata for the regulative principles (and what it might mean for a regulative principle to need a schema).

5.3 Three Regulative Functions of the Ideas and the Focus Imaginarius

On Kant's characterization at A672/B700, the idea of the soul functions (regulatively) to represent a purported substantial underpinning of representations, the one persisting thing to which they all belong as accidents. This is the first of the three regulative functions: the projection of a stable, existent substance, a *thing*.

The idea of the world-whole appears to function differently: not as a stabilizing (projected) substrate but as an impulse, a "push," to find further causes, further (earlier, farther) states of affairs, to remain always dissatisfied with what one has already found (ibid.).[15] This is a second regulative function: impulse to investigation.

It is more difficult to identify the regulative function of the idea of God, despite the fact that Kant discusses this idea most extensively in the Appendix, for it seems just to mirror the regulative principles. When we investigate nature in accord with the (regulatively employed) idea of God, on Kant's view, we consider it "as if" created by a thoroughly rational being, that is, as systematically ordered or (it would seem) in a way governed by the regulative principles (which specify how it is to consider nature as systematically ordered).[16] I suggest that the idea of God might be taken here as furnishing both a mode of interpretation of nature and its (projected) explanatory ground. That is, we are to interpret nature as if created by God, so as if it is thoroughly rational; we are to treat laws or phenomena "as if" they must always have an intelligible reason, all of

[15] So Kant concludes also in his resolution to the antinomies (see, e.g., A520/B548). The regulative function of the idea of the world poses particular questions that warrant more discussion than I can provide here, including how precisely to integrate the resolutions to the antinomies with the Appendix, questions that Kant hints at but does not seem to resolve at A673/B701 and A685/B713. (E.g.: what (if anything) is the regulative role of the concept of freedom? What is to be made of the purportedly contradictory character of the idea of the world itself? Is the world beyond experience in the way that the soul and, especially, God are?)

[16] See especially A700/B728–A701/B729.

which reasons are taken to cohere in one, global understanding of how things are (to be characterized by the ultimate, systematic science).[17] This mode of interpretation – the presumption of the complete, unified rationality of nature – rests on, or perhaps better includes, a (projected) explanation: nature is so, because it was intentionally created by a completely rational agent (or perhaps better: nature may be considered so, if considered "as if" created by such an agent).

According to Kant's descriptions, then, there appear to be three regulative functions of the ideas: projected substance, impulse (to further investigation), and projected unifying rational ground. I suggest, however, that each idea can be seen to function in all three ways (if perhaps with more emphasis on one or another of them). As Kant points out at A683/B711, the idea of the soul encourages the investigator to attempt to explain all psychological powers "as far as possible, as derived from one, unique power," that is, to attempt to find out *more* about the soul, in a way that promotes systematic unity, specifically homogeneity (or finding higher laws or, here, powers). And it does so by being taken as the one (substantial) unifying ground of all psychological states.[18]

In its role as the (projected) unifying ground of nature as completely rational, the idea of God likewise suggests that one should try to know more – in order to approximate to complete knowledge of such a rational nature – and is taken as a substance that underlies (causes) the unity of nature.[19] And though Kant does not explicitly suggest so, it would appear that the idea of the world-whole not only acts as an impulse to further investigation, but also furnishes something at least *akin* to the projected mental substance.[20] For all investigation of further states, causes, and so

[17] We also are thereby purportedly licensed to interpret nature as if arranged for good purposes (see A687/B715), though I shall not discuss this aspect of Kant's view here. In Zuckert 2007, I argue that Kant significantly revised (mostly rejected) this part of his Appendix view in the *CJ*, and omit it for this reason.

[18] For a thorough, if controversial, discussion of the functions Kant allocates to the idea of the soul in empirical psychology, see Dyck 2014: 217–225. Rauscher 2010 suggests that the ideas of soul and world might function as (something like) highest genera and thus as establishing "domains" for the two most general sciences, psychology and physics, a function Kant allocates to ideas in the Architectonic of Pure Reason section of the *CPR*. As Rauscher acknowledges, however, it is not clear why ideas used thus – as domain-identifying genera – would not be considered extremely general empirical concepts, as opposed to noninstantiable a priori ideas, as Kant insists that they are.

[19] Kant indeed refers to the idea of God as a "substratum" of nature at A678/B707.

[20] Despite Kant's denial at A684/B712. Kant takes it that psychological investigation presents particular problems because of the transitory, will-o'-wisp character of mental states, but this (even if correct) does not obviate the likewise projected status of the taken-to-be-there, yet-to-be-found material subject matters of investigation.

forth pushes us "beyond every given experience" (A645/B673), that is, involves aiming to find out about things (events, states of affairs) of which we do not (yet) have experience – for which we therefore have no intuition of a substrate. Here too, it would seem, we project the existence of such a substrate (for the not-yet-discovered empirical particulars), as (I propose) a taken-to-be-already-there part of the world as infinite given whole. Because this idea projects the world as a given *whole*, finally, it may also be taken as a projected unifying ground (of all of nature as interconnected).

Thus, I suggest, the three regulative functions overlap and may be understood to form, together, the regulative role of the ideas. This complex function is summed up, I think, in Kant's famous *focus imaginarius* passage. The ideas, Kant writes there,

> have an excellent and indispensably necessary regulative use, namely that of directing the understanding to a certain goal respecting which the lines of direction of all its rules converge at one point, which, although it is only an idea (*focus imaginarius*) – i.e., a point from which the concepts of the understanding do not really proceed, since it lies entirely outside the bounds of possible experience – nonetheless still serves to obtain for these concepts the greatest unity alongside the greatest extension. Now of course it is from this that there arises the deception, as if these lines of direction were shot out from an object lying outside the field of possible empirical cognition (just as objects are seen behind the surface of a mirror); yet this illusion (which can be prevented from deceiving) is nevertheless indispensably necessary if besides the objects before our eyes we want to see those that lie far in the background, i.e., when, in our case, the understanding wants to go beyond every given experience (beyond this part of the whole of possible experience), and hence wants to take the measure of its greatest possible and uttermost extension. (A644/B672–A645/B673)

In this comparison of the ideas with object-images in mirrors, Kant refers to all three regulative functions. The ideas are to push us "beyond every given experience" so as to give the understanding a greater "extension" (the impulse function). But they also are "as if" *objects* (the projected substance function). *And* the focal point is to ground a systematic unity among particular objects or judgments: the understanding's "lines of direction" are united in aiming toward it – or, in terms of another metaphor in the Appendix, all the visible objects are arrayed across a single horizon (unifying ground function).[21]

[21] A658/B686–A659/B687; Kant uses the horizon as an image not for the regulative use of the ideas but for the results of investigation in accord with all three regulative principles.

In less imagistic language, Kant is proposing that the ideas of reason (as "focus imaginarius") comprise representations of something that is (projected as) *there* (a purported object), beyond the current state of knowledge, and either the source of, part of, or identical to a rationally unified world.[22] As I shall now suggest, this understanding of the ideas' role (as *focus imaginarius*) both reraises my overarching question – why could one need the ideas of reason in empirical investigation? – and can help to answer it.

5.4 Schemata for the Regulative Principles

The preceding discussion can, I suggest first, answer a subsidiary part of my question, namely, what specifically the ideas are supposed to contribute (beyond the regulative principles) to guiding empirical investigation. For it would appear that the ideas represent *preconditions* for investigation in terms of the regulative principles. In order to pursue greater systematic unity of empirical knowledge – to find more universal, more specific, and more continuous laws – one must presuppose that there *is* something out there to be so investigated, a something that lies *beyond* our current state of knowledge *and* that will be part of a rationally ordered world. One must, in short, have hope that there is something to be found, and to do so, one needs to have the sense that investigation is *about* something. Such hopeful presuppositions are distinct from the regulative principles: they not only have different content, but also are a different, one might say, "objectival"[23] *sort* of representation. They are not directives ("look for this"), as are the regulative principles, but rather present (purported) objects, to be governed by the laws we hope to discover. And the ideas – as representations that seem to present us with particular objects – are arguably well suited to perform this role as preconditions or (perhaps better) heuristic posits.

I suggest, in other words, that the ideas function here similarly to (if also more weakly than) the postulates of practical reason, that is, the morally

[22] Perhaps, then, there is ultimately just one *focus imaginarius* – an intelligible kingdom of nature, a systematically ordered world, rationally created by God, containing both mental and material substances – that is the schema (singular) for the aspiration to systematic science expressed in the regulative principles.

[23] I use this invented term to convey that the representation (idea) seems to present us with an object (rather than a rule or class for combining or classifying objects), but without any of the connotations of success or validity carried by "objective." Insofar as the ideas are thus "objectival," they are similar to sensible representations (which in fact, not just purportedly, present particular objects), and thus could be understood as analogous to schemata proper (which are sensible).

required beliefs that the soul is immortal and that God exists. Here too (I am suggesting) the ideas function as theoretical assumptions – as "objectival" representations, in the terms I've used – presupposed by a "practical" endeavor, in this case a theoretical practice (as it were) of ongoing scientific investigation.[24] On Kant's view, we are of course not enjoined as a matter of *duty* to aim at systematic empirical science (as we are morally enjoined to seek the highest good). Nor, I suggest correspondingly, are we required to *believe* that the systematically ordered, infinite intelligible world *exists* (as we are required to believe that God exists for the purposes of morality, on Kant's view). Rather and more weakly, we need to have a sense, a projected image of such a world, as something "out there" that might be investigated (profitably). We need (what I call) an optimistic placeholder, a stand-in for the "I know not what" we hope to find in investigation. Or, in Kant's own terms, "reason cannot think this systematic unity in any other way than by giving its idea an object, which, however, cannot be given through any experience; for experience never gives an example of perfect systematic unity" (A681/B709).

Yet one might still wonder whether one needs the *ideas* specifically here – why must one posit God, the infinite world-whole, the soul? Why not take it that one is trying to find out more (and more systematically) concerning the objects one has *already* identified in accord with *empirical* concepts? This would reflect the facts, after all: one is trying to know not about God or the soul, but rather about *nature*, that which is presented in experience and represented in accord with empirical concepts. Or, failing that, why not some more empirical (and more contentful) posit, like the "ideas" of pure earth, pure air, pure water that Kant himself mentions as heuristic concepts that might help in finding chemical laws (A646/B674)? Or, again, why not the more generic optimistic presupposition just sketched – namely, that there is something out there, beyond what we (currently) know, that is rationally ordered?

The concept of a *focus imaginarius* may help, in other words, to explain why a further objectival representation might be used to supplement the regulative principles, but it also raises again the questions with which I began. For in his description of the *focus imaginarius*, Kant emphasizes the non-sensible beyondness of the ideas – precisely that which renders them incapable of serving in any straightforward sense as schemata, as

[24] Kant characterizes a postulate of practical reason as "a *theoretical* presupposition, though one not demonstrable as such, insofar as it is attached inseparably to an a priori unconditionally valid *practical* law" (*CPrR* 5:122).

mediations between a priori principles and sensible experience. Why should a *focus imaginarius*, a goal forever beyond reach, outside experience, help to employ the regulative principles *within* experience?

Kant suggests, in the passage just quoted from A681/B709, that the ideas are "perfect" as no empirical object can be – and thus they might serve as aspirational object in a stronger sense, namely, as ideal standard. This is, I think, right in some sense (as I shall discuss in a moment), but again one might ask: why set a standard of an entirely different kind from whatever results one will in fact attain, even the very best ones? Systematic unity – and the regulative principles thereof – may well constitute a not-quite-reachable goal for empirical science, and so a standard according to which any current empirical science might be judged. But no empirical science will ever comprise anything like, even "asymptotically" approaching, rational knowledge of the soul, infinite world, or God.

To respond to these questions, I return to a point made briefly above: that the regulative principles give only very general, vague guidance concerning how to investigate nature. This vagueness, as I noted, explains their status as *merely* regulative – "regulative" used in a hedging sense, to mean that the principles do *not* tell us anything about how objects might be. But it is also, I now add, crucial to their legitimacy *as* a priori principles, specifically as regulative principles in the more literal *guiding* sense of "regulative." These principles are, Kant argues, legitimate precisely *because* they provide mere, indeterminate guidance as to what to look for within the empirical (A509/B537). They do not pretend to establish anything about objects a priori, and so do not impinge on properly *empirical* investigation of nature.[25] More strongly, and positively, the regulative principles would not guide us toward *greater* unity of empirical science, *further* extension of experience if they were taken directly or determinately to apply to any empirical object, any occurrent state of empirical science.[26] In that case, we would take it that we had accomplished ideal science, satisfied rational demands already, a problematic stance akin to that which Kant names "lazy reason" (A689/B717). Unlike "lazy reason," such empirical complacency would arise not from a purported a priori grasp of objects as intended by God to be as they are, but rather from the view that current empirical science explains everything that

[25] They are also, of course, legitimate because their aim – full rational systematicity – reflects the nature and aspirations of reason.

[26] See A680/B708 where Kant moves from a negative, hedging characterization of the vagueness of the regulative principles to a positive, guiding one (they do not determine objects, but do open up new paths).

there is to be explained.[27] But like lazy reason, such complacency would encourage neglect of inquiry. In sum, to function as regulative for empirical investigation, a principle *must not* be taken as directly to be applied to any empirical object or (more properly) to any set of empirical judgments, laws, or concepts.

If this is right, the schema of a regulative principle – a representation that aids in its use or application to particulars – cannot perform the ordinary function of a schema either. The appropriate use of such principles precisely does not require some representation to mediate, to enable their fit, their direct application to sensible particulars. Rather, I propose, here the schema must always defer such fit, obstruct any sense of seamless application, in order to prevent empirical complacency. It must be, that is, an idea – a *focus imaginarius*, a (purported) object standing in for what might be found, but that in fact will *never* be the subject matter of empirical scientific knowledge.

Let me spell out this proposal somewhat, by comparing the ideas to the other possible objectival representations (suggested above) that might be used in empirical scientific investigation. First, one must grant, of course, that scientists aim to find out more about granite or gold or trees, and so forth – that is, of course scientists take objects as described by empirical concepts to be the things they are aiming to investigate. But somehow – behind, above, beyond that – one also needs to think that these concepts might be wrong, might require (even drastic) revision: the higher genus concept of mechanically ordered matter revised Aristotelian empirical concepts of kinds, the specification of "jade" required recognition that in fact it was not a genus, and so forth. Thus the empirical objects to be investigated cannot be strictly, nonrevisably *fixed* by empirical concepts – this would be, precisely, a form of (what I am calling) empirical complacency.[28] Arguably, then, the investigator needs a placeholder to refer to what one might find that is neither identical to any empirical concepts we have nor determinate a priori in such a way that would shut down or predetermine the results of empirical investigation. The ideas, I am suggesting, function as just such (in principle and always) indeterminate placeholders. They are useful for empirical investigation precisely because

[27] Kant claims that the term "lazy reason" can describe "any principle that makes one regard his investigation into nature ... as absolutely complete" (A689/B717), thus including what I am calling empirical complacency, but I think it helpful to mark different sources of such laziness terminologically.

[28] Kant does not explicitly so argue, but does refer to the regulative function of rational ideas as needed to "cultivate *and correct*" the claims of the understanding (A671/B699, my emphasis), which presumably includes empirical conceptualization.

they are both nonempirical and almost empty, unknowable in principle. Hence, I think, Kant's remark at A681/B709 – following his claim, quoted above, that reason thinks systematic unity by positing a priori an object for it – that this idea "can never be obstructive" but only "conducive" to the "empirical cognition of the understanding": in its unknowability and emptiness, the idea does not obstruct actual empirical investigation.

The ideas function, therefore, quite differently from "ideas" like those of pure earth, air, and water – even though these latter ideas are also formulated by reason and helpful in empirical investigation (and also discussed in the Appendix, though in the first, rather than second, section). As Kant notes, insofar as we conceive the elements (earth, air, water) as absolutely pure, these representations are formed by reason or are (one might say) relatively a priori. They are (in our terms) idealizations both in being not immediately experienced as such (in experience they are impure, i.e., found in combination) and in being rationally "better" (simpler, more easily understood, more suitable as a subject of a universal law). The ideas of the pure elements are also, like the ideas proper, posited-object-representations – representations of purported objects that are to be governed by the laws we hope to find. Importantly, unlike the ideas of reason, however, the objects of these ideas are not necessarily or irrevocably beyond experience. Indeed, Kant refers to pure earth, air, and water as "natural causes" (A646/B674) and takes them precisely to be (proposed) explanations for how natural objects behave. For they are, Kant claims, employed in the "hypothetical" use of reason, where a rule is "assumed only *problematically*" to see if "particular cases, which are all certain . . . flow from it," in which case the rule is "inferred" to be universal (A646/B674–A647/B675).[29] Thus, by contrast to the "as if" regulative consideration of the ideas proper, and Kant's stress on the illegitimacy of deriving anything empirical from them, we aim here precisely to derive actual empirical states of affairs from such hypothesized causes.[30] These

[29] Kant's description of such hypotheses as "regulative" appears largely to be hedging, to indicate that we do not, strictly, *know* the "rule" to be universal, because we do not know that *all* of its consequences hold (see A647/B675). But they also arguably function (at least locally) as guides (i.e., positively regulative) in the search for more laws (whether more specific than Newtonian mechanics or more general than observations concerning freezing temperatures, say) or for their own dis/confirmation (as suggested in Rohlf 2014).

[30] With Rauscher 2010, and contra Grier 2001: 297–301, whose view is otherwise quite consonant with my proposals here (I believe). Kant confuses matters, I think, in suggesting that the ideas of reason are also posited "problematically" (A681/B709): they are not, indeed, posited assertorically, but they are also not posited in a way that might be converted to an assertoric claim either; they are (as it were) permanently problematic.

hypothesized ideas *do* impinge on the empirical, then, either (if unsuccessful) as disproven hypotheses or (if successful) as new empirical explanations or conceptualizations. They are tools indeed in rational systematization, but local and temporary ones: they do not keep open the possibility that there is always more and perhaps radically different systematization to be done.

If we do need to keep open this possibility – if we need, as Kant puts it, to "open up new paths into the infinite (the undetermined)" (A680/B708) – then perhaps we need to have a sense that there *might be* objects different from – beyond – natural objects as currently described by empirical science or even by our wildest hypotheses.[31] And so finally, in suggesting that we need the ideas to function as a *focus imaginarius*, Kant implies that this propositional belief (that objects might be different from what empirical science now holds) is, somehow, insufficient. Kant does argue in the first section of the Appendix that in investigation guided by the regulative principles, we must presuppose something like it, namely, the "transcendental principle . . . through which . . . a systematic unity, as pertaining to the object itself, is assumed a priori as necessary" (A650/B678). In attempting to find higher or lower empirical laws, that is, we presuppose that natural objects *are* such as to correspond to those systematizing aspirations, are governed by such (as yet unformulated) laws. Though the meaning and status of this presupposition are again subjects of much discussion,[32] broadly speaking it seems to articulate both *what* investigation according to the regulative principles aims to accomplish – knowledge of more universal or specific characteristics *of objects* (not just logical (genus/species) rearrangement of our concepts) – and the (as it were) hopeful belief required for this endeavor, namely, that there are such characteristics in nature to be found.

This transcendental presupposition thus has some of the objectival and optimistic purport that, I have suggested, is contributed to scientific investigation by the ideas, regulatively employed. I would suggest, however, that in its focus on nature, it makes both stronger and weaker claims than do the ideas, and thus plays a different (but complementary) function in scientific investigation on Kant's view. On the one hand, this presupposition asserts that the higher or lower laws will be *true of* some (natural)

[31] These are themes I have touched on concerning reflective judgment and its principle of purposiveness in Zuckert 2007, differently, but I believe complementarily, to the present proposals.

[32] See, e.g., Guyer 2003b and Chapters 3 and 4 in this book.

objects – whereas the ideas *merely* present images of objects, objects "in the idea" (as Kant puts it at A670/B698), reference points or placeholders for investigation, about which nothing ever really will be known. The presupposition thus expresses a stronger claim, a rational assumption for investigation not furnished by the ideas.

On the other hand – and more importantly here – the presupposition more weakly supports the regulative principles' *demand* to proceed further in investigation. It does state that we should be confident in trying to find out more about objects, that we may rationally assume that there is systematic order to be found in nature. But it does not itself suggest that things might be different from how current empirical science has it – perhaps current science reflects the degree to which nature really is systematic. Or, in other words, insofar as the regulative principles articulate a goal – systematic unity of science – the presupposition supports them in affirming the realizability of that goal. But the ideas support, as it were, its character as a *goal*, its demandingness. For the transcendental presupposition does not include any sense of that which is beyond current experience, current scientific hypotheses, beyond nature as we now find it. And even if it were to do so – if one were to restate it in the terms suggested above, namely, to include explicitly the presupposition that (systematically unified and rational) objects might be *quite different* from what our empirical science has so far dreamed of – by itself it might, I suggest, appear to be *just* an empty presupposition, a *mere* idle thought that could be turned aside by arguing that it has little evidence in its favor, little reason to consider it seriously, compared with the vast success of (any current) empirical science, the vast evidence we have of its conformity to natural objects.[33] It would not, that is, stand much in the way of empirical complacency.

The ideas present, instead, specific objects – soul, infinite world, and God – and do so not arbitrarily, not as mere idle thought, but (if Kant is correct) because they express central demands of reason. The objects of these ideas are, moreover, presented as rationally superior – more consonant with the demands of reason (indeed they are hypostatizations of those demands) – than are natural objects, even those understood in accord with the best empirical science. The idea of the soul – in its simplicity and pervasiveness in all thought – is an idea of an object able to be understood

[33] These concerns (about the "idle thought") are somewhat similar – if transposed to the realm of empirical science, rather than philosophical reflection – to objections Langton 1998 raises against the "anodyne interpretation" of transcendental idealism.

through-and-through in its most fundamental and universal character. The idea of God presents an object as thoroughly rationally determined, all its properties both necessary to it and related to one another for reasons, down to its very existence (as purported necessary being).[34] And of course the idea of the infinite world-whole purports to encompass everything.[35]

By contrast, as Kant reminds us in the Dialectic, nothing empirical counts as an absolute "boundary," and thus we will never (definitively) attain knowledge of the whole (A517/B545). No empirical concept allows its object to be known in its simplicity through-and-through but is rather a discursive and inductively formed universal. And (Kant claims) "no species [or empirical concept thereof may] be regarded as in itself the lowest" (A655/B683), thus no particular can be conceptually grasped as thoroughly determined; nor is there any empirical knowledge of (nonhypothetical) necessary existence (see, e.g., A559/B587).[36] In short, the ideas contain – if sketchily, if imaginarily – a specific suggestion that the world might be known otherwise, and known better.

The ideas serve, then, to resist empirical complacency – satisfaction with current empirical science – by providing a specific reference point to mark what has not been accomplished. Thereby they also resist *empiricist* complacency, namely, the (global, philosophical) view that empirical scientific knowledge is the highest or only conceivable form of knowledge. Kant holds, that is, that we improve empirical science by attempting to approximate to systematic knowledge, a standard that lies beyond its current state and arises from other origins (not experience, nor even a priori principles of its possibility, but pure reason). Thereby we make empirical knowledge more comprehensive, more responsive to empirical determinacy, and more rationally unified (see A677/B705). I have proposed that the ideas contribute to this enterprise by serving as optimistic placeholders in scientific investigation. But the ideas play a further regulative role with respect to empirical science, I suggest now: of limit

[34] I take it that this is Kant's explanation of the origin of the concept of God as Ideal (see A576/B604–A580/B608), combined with the rational need for an ultimate, necessary ground of existence (see, e.g., A584/B612) – though on Kant's view, of course, we cannot justify this combination, i.e., show why the *ens realissimum* necessarily exists (see, e.g., A612/B640).

[35] *Possibly* this maximally rational content for the ideas could link their function as schemata for the regulative principles to Kant's other, puzzling reference, in the first section of the Appendix, to using an idea – "of the *maximum* of division and unification of the understanding's cognition in one principle" – as an "analogue" to a schema for systematic unity (A665/B693). This latter idea may, however, be the rational idea of systematic unity (as aim and standard for scientific investigation) itself, and not equivalent to the ideas of reason narrowly, on which I focus here.

[36] Kant of course holds that we do not understand the (purported) necessity of God's existence either; see, e.g., A679/B707.

to its pretensions (to absoluteness, to exhaustiveness, to rational comprehensibility). Kant suggests both roles in his single explicit reference to the ideas in his later *CJ* discussion of empirical investigation. He repeats there the now-familiar claim that the ideas "guide" the understanding "in the contemplation of nature in accordance with a principle of a completeness to which it can never attain" – that is, I take it, that of systematic unity. He adds that they "restrain the worrisome pretensions of the understanding, as if (in virtue of being able to furnish a priori the conditions of the possibility of all things that it can cognize) it has thereby also confined the possibility of all things in general within these boundaries" (*CJ* 5:167–168). In providing perfect images of perfect knowledge, the ideas mark what will not be achieved by the understanding – by any empirical science, even the most systematic.

This final point is why I have been referring to the ideas as *nearly* empty of content. For the ideas must have *some* content in order to function as nonarbitrary ideal (unattainable) standards (or indeed as *optimistic* placeholders).[37] On the other hand, this qualification ("nearly") is less significant than may appear. For, I suggest, it is (again) their near emptiness, the thinness of their content, that allows the ideas to give the impression of rational superiority, of greater knowability of their objects. In fact we know nothing about the immaterial soul, God, or the infinite world-whole on Kant's view: not only have we no way of encountering those objects, of checking whether claims we make about them are indeed so, but also we have little conception of what we should claim, of what sort of property we should attribute to them, in the first place.[38] Yet the thinness of their content can be (mis)perceived as rational simplicity and completeness, as that which is, thereby, thoroughly understood – only so can they give us a mirage-image of knowledge superior to the only sort of knowledge we in fact have.

5.5 Conclusion

I have argued that on a Kantian view the ideas serve as optimistic placeholders in empirical scientific investigation: they stand in for

[37] Kant does state that we posit "a something about which we have no concept at all of how it is in itself" in using an idea as schema of the regulative principles (A674/B702). This is a strong and (I think) somewhat inaccurate description, since we do have *some* (rather thin) "concept" of the object of the ideas, though of course we cannot be confident that it represents how they are "in themselves."

[38] See A676/B704–A677/B705. More generally, this appears to be the purport of many arguments in the Paralogisms and the Ideal chapter: all contentful claims made about these purported objects illicitly borrow their "materials" from experience.

whichever objects, however described, we will have discovered in (future, successful, systematizing) investigation. They can perform this role because they are beyond sensible experience and empirical conceptualization and because, as mere projections of reason's demands for unity and totality, they are at once nearly empty and yet not arbitrary. The ideas thereby leave empirical investigation free to find what might be there to be found, yet also provide an image of rationally ideal knowledge, beyond any possible empirical knowledge. And so they not only encourage, but also humble, in a way reminiscent of Kant's most general, transcendental idealist conclusions. But transcendental idealism is, arguably, also the precondition for the constellation of regulative functions Kant allocates to the ideas: if they are to serve as nonarbitrary, optimistic presentations of potential objects of knowledge, from which, nonetheless, one is never to derive any actual knowledge – if one is to resist both empirical complacency and lazy reason – then it must be the case that we can *think* such objects, but never *cognize* them.

Acknowledgments

I am grateful to Steven Crowell, Andrew Cutrofello, Brian Mellovitz, and participants at the *Kant, the Laws of the Life Sciences, and the Lawfulness of Nature* conference at the University of Edinburgh in June 2015, for discussion of the issues raised in this chapter, and to the editors for insightful comments on an earlier draft. All errors are my own.

CHAPTER 6

Kant's Transcendental Principle of Purposiveness and the "Maxim of the Lawfulness of Empirical Laws"

Thomas Teufel

6.1 Introduction

Kant presents his mature discussion of the necessity of empirical laws of nature in the Introduction to the *Critique of the Power of Judgment.* The nature of Kant's answer to Hume – specifically, the answer to the question of whether critical philosophy can counter regularity theories of causation with a form of empirical (causal) realism – largely depends on how one interprets Kant's position there. Prior to the third *Critique,* Kant had discussed the necessity of empirical laws of nature in the Analogies of Experience[1] and, notably, in the Postulates of Empirical Thinking in the *Critique of Pure Reason.* On his view in the Postulates, to attribute necessity to objective sequences of events – specifically, to Humean constant conjunctions – is to judge that those sequences "accord"[2] with transcendental philosophy's formal, spatiotemporal, and categorial framework. In the words of a recent commentator, such a judgment "transform[s]" (Friedman 2014: 532) an otherwise merely empirical regularity into a causal law of nature. But the precise nature of the transformation envisioned – which proceeds by way of a postulation, namely, a subjective reflection on an already objectively determined sequence – remains elusive in the *Critique of Pure Reason.* Consequently, just what the "material necessity in existence" (A226/B279) thus attributed to empirical regularities amounts to remains elusive as well.

Kant picks up the thread in the Introduction to the *Critique of the Power of Judgment.* The central novelty in Kant's treatment of the necessity of empirical laws of nature in the third *Critique* is that Kant now ties it to the

[1] "But this rule is always to be found in the perception of that which happens, and it makes the order of perceptions that follow one another (in the apprehension of this appearance) *necessary*" (A193/B238).

[2] "That whose connection with the actual is determined in accordance with general conditions of experience is (exists) *necessarily*" (A218/B265).

notion of nature's "lawful unity" or "systematic unity" (*CJ* 5:184). Specifically, in §V of the Introduction to the third *Critique*, Kant addresses the necessity of empirical laws of nature as part of his transcendental deduction of the principle of this systematic unity, "the principle of the formal purposiveness of nature" (*CJ* 5:181). The transcendental status of the principles of nature's systematic unity had, to be sure, already played an important role in the Appendix to the Transcendental Dialectic of the *Critique of Pure Reason*, specifically, in Kant's attempt to ground the subjective necessity of heuristic maxims of scientific inquiry.[3] But Kant's account of such principles in the Appendix did not take nature's systematicity to have bearing on the question of the necessity of empirical laws of nature (not to mention that Kant struggled, unsuccessfully, to explain the sense in which these principles could be both noncategorial and yet transcendental).[4] In the *Critique of the Power of Judgment*, then, two previously separate strands of argument – one concerning the necessity of empirical laws, the other concerning the systematicity of nature – are woven into a complex new tapestry.

Largely as a result of the tightness of Kant's weaving, commentators typically assume that the convergence of these two strands of argument means that both now form a unified account.[5] The assumption is that Kant's transcendental principle of nature's purposiveness in the *Critique of the Power of Judgment* is a principle *concerned with* the necessity of empirical laws of nature – that is, a principle whose cognitive role consists in casting empirical laws of nature as necessary and whose transcendental-philosophical justification stems from the nature and success of its discharge of that role.

This view, however, oversimplifies the matter. In his transcendental deduction of the principle of nature's purposiveness in the *Critique of the Power of Judgment*, Kant jointly tallies what one might call the transcendental and the epistemic consequence of a counterfactually presumed *absence* of the new principle of nature's purposiveness. That is to say, Kant contemplates what our cognitive situation would be like if the transcendental principle of nature's purposiveness did not obtain. Kant's account of the envisioned transcendental consequence forms the heart of his deduction, as it helps establish the transcendental necessity of the principle of nature's purposiveness – but it does not concern the necessity of empirical laws of nature. Conversely, Kant's account of

[3] See A648/B676, A662/B690. [4] See Guyer 1990: 28; Kemp-Smith 1992: 552; Grier 1997: 5.
[5] See Allison 2001: 38; Guyer 2003b: 87ff.

the envisioned epistemic consequence does concern the necessity of empirical laws of nature – but it does not contribute to establishing the transcendental necessity of the principle of nature's purposiveness.

The *transcendental* consequence of a counterfactually presumed absence of the principle of nature's purposiveness is a form of thoroughgoing cognitive chaos, which Kant invokes twice in §V of the Introduction but whose depths he does not fully probe until §§76–77 of the third *Critique*. Only in this latter (justly famous) form does the envisioned consequence lend full support to the principle's transcendental status.[6] The problem addressed in §§76–77 is the threat of a fundamental incompatibility between sensible particularity and conceptual universality (pure or empirical) – a threat neither seen clearly nor addressed fully in the *Critique of Pure Reason*[7] – that would undermine the possibility of any human cognition at all, were it not for the (accordingly) a priori and transcendentally necessary principle of nature's purposiveness.

The *epistemic* consequence of a counterfactually presumed absence of the principle of nature's purposiveness, which will be my main concern here, is less transcendental-philosophically dire, although spelling it out presents its own rhetorical challenges. After all, if the foregoing is correct, then, in the absence of a principle of nature's purposiveness, there would be no possibility of any cognition at all. A fortiori, no *further* epistemic consequences could possibly arise from that absence. Kant's point, as I see it, is that in the *presence* of the principle of nature's purposiveness we are entitled to something that we could not possibly be entitled to in the counterfactually presumed *absence* of that principle, even if that absence did not have the serious transcendental-philosophical ramifications Kant establishes in §§76–77. Kant's point is that in the presence of the principle of nature's purposiveness we are – but in the counterfactually presumed absence of that principle could not possibly be – entitled to consider well-established regularities in nature as necessary and, hence, as laws.

Kant is clear that the lack of such entitlement would only frustrate a subjective and contingent (or at any rate not a transcendentally necessary)

[6] Thus, Kant's transcendental deduction is more protracted than its official presentation in the introduction would suggest (see Allison 2001: 38; Teufel 2012: 318–324). This is, no doubt, a central reason for the continued scholarly uncertainty concerning not only the success but even the presence of a transcendental deduction of the principle of purposiveness in the third *Critique* (see Brandt 1989: 187).

[7] See Teufel 2012: 313, 315–318.

end of the understanding, namely, to hold consistent beliefs (see Section 6.6, below). Accordingly, the threatened lack of entitlement cannot itself serve as a premise in an argument to support the transcendental necessity of the principle of nature's purposiveness. Still, the envisioned epistemic consequence does reveal that the principle of nature's purposiveness has at least two important cognitive functions. In addition to its fully transcendentally necessary role in warding off cognitive chaos, it serves as the rationale behind an instrumental maxim that warrants attribution of a material form of necessity to empirical laws of nature. I will call the maxim thus backed by – and established alongside – the transcendental principle of nature's purposiveness the "Maxim of the Lawfulness of Empirical Laws."

The maxim governing attributions of necessity to empirical regularities is supported by the transcendental principle of nature's purposiveness. Kant's argument in favor of this maxim thus presupposes, but is not itself part of, Kant's argument for the transcendental status of the principle of purposiveness. When this is understood, the nature of the philosophical problem that the necessity of empirical laws of nature presents for Kant – as well as the philosophical significance of Kant's solution to that problem, hence, the overall shape of his answer to Hume in the third *Critique* – comes into proper focus.

The philosophical problem concerning our attribution of necessity to candidate empirical regularities is strictly one of showing that we are *entitled* to the assumption that those regularities are necessary, hence, that our conceptualizations of candidate regularities express genuine laws of nature. We need neither *know* that those empirical regularities are in fact necessary, since that is in principle unknowable, nor do we require a principle that separately prompts or exhorts us to *make* the assumption that those regularities are necessary, since that assumption is analytically entailed by their nature as causal generalizations.

Claiming entitlement to a mere assumption of necessity (let alone an assumption the understanding makes by default) may, however, seem like a peculiarly weak defense of empirical (causal) realism in the face of the skepticism of regularity theories. Can this possibly be Kant's vaunted answer to Hume? Here it must be kept in mind that the transcendental principle of nature's purposiveness also supports a set of maxims of scientific inquiry, which, in combination with the entitlement in question, suggest a powerful iterative procedure by which to claim warrant for the presumed material necessity not just of any but of an ever expanding and ever methodically refined set of empirical regularities.

I begin my account with a discussion of the transcendental deduction of the principle of nature's purposiveness in the published introduction to the *Critique of the Power of Judgment*, focusing in particular on the distinction between, on the one hand, the problem of nature's systematic unity and, on the other hand, the problem of the necessity of empirical laws of nature (Section 6.2, below). This is followed by a closer look at the basic *structure* of the principle of nature's purposiveness, as a *regulative, heautonomous demand for a presupposition of purposiveness* (Section 6.3, below), as well as a discussion of the bearing this structure has on the nature of that principle's *claim*, as minimalist and not maximalist in ambition (Section 6.4, below). Understanding the principle's structure and claim, in turn, helps establish the nature of the *methodological* maxims Kant derives from it (Section 6.5, below). This, in turn, helps explain the nature of the maxim of the lawfulness of empirical laws (Section 6.6, below) and, in particular, the nature of the support that maxim receives from the transcendental principle of nature's purposiveness (Section 6.7, below).

6.2 Kant's Transcendental Deduction of the "Principle of Nature's Purposiveness"

Kant begins his transcendental deduction of the principle of nature's purposiveness by observing that such a principle "shines through" (*CJ* 5:182) in our methodological maxims of scientific inquiry. Kant's suggestion is that heuristic maxims, like Occam's razor or the principle of nature's gradualism ("*natura non facit saltus*"), presuppose an underlying systematic unity of nature that is neither merely empirical (because the presumed unity is not itself observable) nor merely psychological or subjective (because the presumed nonobservable unity is nevertheless supposed to be objective). But since these methodological maxims do not themselves serve an indispensable cognitive end, they may suggest, but they cannot in turn help establish, that principle's transcendental status. For the transcendental deduction proper, therefore, Kant turns to a different line of argument, which revolves around the unity of our empirical generalizations.[8]

Initially, Kant's idea appears to be that if we imagined – contrary to the presumed transcendental fact – that a principle of nature's purposiveness were *not* part of our cognitive makeup, then ordinary empirical cognitions might be subject to a striking degree of disunity. In the absence of a

[8] See *CJ* 5:183–186.

principle of nature's systematic unity, we would still be able to attain a ground level of empirical object-consciousness by means of the cognitive apparatus established in the Transcendental Aesthetic and the Transcendental Analytic of the *Critique of Pure Reason*; thus, our finite minds would enjoy empirical cognitions of causally interacting substances in space and time. But even if they were thus fully determined by the transcendental conditions of the possibility of experience, Kant suggests that the resulting empirical cognitions might well be merely fleeting, episodic, and impossible to integrate into a stable system of cognition.

Specifically, while the cognitive episodes in question would still support empirical generalizations and, hence, judgments of lawful regularities, there is no transcendental-logical assurance against the possibility of an infinite proliferation of fully distinct, often contradictory, yet nevertheless empirically equivalent generalizations. Each of these generalizations would, qua causal judgment, invoke a causal law, but none of these generalizations could justify that law's implicit claim to empirical necessity since none could lay claim to capturing empirical reality uniquely.

The problem Kant points to here is not simply one concerning the possibility of a plethora of changing conditions and intruding factors that might hamper the formulation of truly universal and necessary laws in the special sciences – factors that can be addressed by *ceteris paribus* clauses. The scenario Kant envisions is one in which *ceteris paribus* clauses are themselves not justifiable since they are not entailed by the categories or the system of principles and since no other transcendental-logical justifying grounds are (in the presumed absence of a principle of nature's systematic unity) on offer. It is a scenario in which *all* else might be equal in the initial conditions and, yet, a different causal outcome would still remain entirely possible and, indeed, fully plausible. In Henry Allison's formulation, it is a scenario "in which something like Hume's 'uniformity principle' does not hold" (2001: 38). Kant puts the point this way:

> Now, aside from that formal time-determination [author's note: aside from the category of causality and in the absence of a principle of nature's systematic unity], the objects of empirical cognition are, as far as one can judge *a priori*, still determinable in many ways, such that specifically distinct natures, aside from what they have in common as belonging to nature in general, can still be causes in an infinite variety of ways; and each of these ways must (according to the concept of a cause as such) have its rule, which is a law, and hence is accompanied by necessity, even as we cannot have insight into this necessity according to the constitution and limits of our cognitive capacities. (*CJ* 5:183, author's translation)

"As far as one can judge *a priori*," our cognitive interactions with the world of outer experience could issue in wildly inconsistent empirical generalizations with associated wildly inconsistent claims to empirical necessity. As a result, "no thoroughgoing interconnection of empirical cognitions into a whole of experience would occur" (*CJ* 5:183).

This, to be sure, is a problem for transcendental philosophy more than for ordinary empirical consciousness or natural science, both of which do arrive at more or less stable (if improving) representations of empirical reality. Kant's complaint, then, seems to be that, from the vantage point of the *Critique of Pure Reason*, this happy fact about ordinary empirical and scientific knowledge can be little more than a lucky coincidence – and not at all one determined (or at the very least guided) by the sophisticated transcendental principles that (there) describe our cognitive relation to the world. It is entirely consistent with Kant's mature theory of object-cognition in the *Critique of Pure Reason* that our cognitive reality might be one of, in Allison's words, thoroughgoing "empirical chaos" (2001: 38).

Kant concludes that, confronted with this possibility, the power of judgment "must . . . think nature on behalf of our power of knowledge according to a principle of purposiveness" (*CJ* 5:184). He proceeds to consider the notion of purposiveness in question a "transcendental concept" and refers to this whole line of thought as a "deduction of the present concept [of purposiveness] and of the necessity to regard it as a transcendental principle of knowledge" (ibid.).

Kant's transcendental deduction of the principle of nature's purposiveness, then, appears to be this: if a transcendental principle of nature's systematic unity were not part of our cognitive makeup, then there would be no assurance of our ability to consider empirical laws of nature as necessary and, a fortiori, to integrate them into a coherent system of thought. But we do consider empirical laws as necessary and, hence, integrate them into a (more or less) coherent system of thought. Accordingly, a transcendental principle of nature's systematic unity must be part of our cognitive makeup.

This, of course, is a non sequitur. The threat of chaos envisioned is ultimately rooted in ontology, while the transcendental-philosophical solution proffered is purely epistemic. Unfortunately, there simply is no transcendental-philosophical remedy for an inherently uncooperative world. If the world were indeed fundamentally changing and unstable so as to allow us to arrive only at conflicting empirical generalizations, then we would remain unable to consider the *in*-constant conjunctions we thus come across as truly necessary laws and to integrate them into a coherent system of thought – even *with* the additional (accordingly, superfluous) transcendental-philosophical assurance of nature's underlying systematic unity. Conversely, if the world were fundamentally stable so as to allow us to arrive at interconnected empirical generalizations, then we would be able to consider the *constant* conjunctions we come across as necessary laws and to integrate them into a coherent system of thought even *without* the additional (accordingly, superfluous) transcendental-philosophical assurance of nature's underlying systematic unity.

This helps clarify the relation between, on the one hand, the topic of nature's systematic unity and, on the other hand, the topic of the necessity of empirical laws of nature. Whatever threat of chaos (if any) may ultimately ground the transcendental legitimacy of a principle of nature's systematic unity, that threat is not the problem of empirical chaos Allison mentions[9] or the associated problem of a coherent system of empirical laws. These problems are impervious to transcendental address. The question of the transcendental justification of the principle of nature's purposiveness thus remains as yet unresolved since its proper transcendental ground remains as yet unidentified. Still, Kant's placement of the problem of the necessity of empirical laws in the context of a counterfactually presumed absence of the transcendental principle of purposiveness appears to be more than merely a false start on the road to that principle's eventual transcendental justification. As we will see, a transcendentally warranted principle of nature's purposiveness would indeed support the claim to necessity implicit in our formulation of empirical laws of nature even if that cognitive function does not, in turn, supply the principle's transcendental warrant. Kant's placement of the problem of the necessity of empirical laws in the context of a counterfactually presumed absence of the transcendental principle of purposiveness thus suggests that in addition to a transcendental consequence, that counterfactually presumed absence would have an independent epistemic consequence as well, one related to our ability to consider candidate empirical generalizations as laws.

6.3 The *Structure* of the Principle of Nature's Purposiveness

Before investigating the suggested epistemic consequence of a counterfactually presumed absence of the transcendental principle of nature's

[9] Indeed, Allison himself seems at times skeptical that this type of chaos alone can capture the problem Kant hopes to address; see Allison 2001: 39. For discussion, see Teufel 2012: 309–313.

purposiveness (see Section 6.6, below), it is necessary to present the basic *structure* and *claim* of that principle. It should be noted, however, that understanding the structure and claim of the principle of nature's purposiveness still falls short of answering the question of the transcendental *warrant* of Kant's proposed solution to the problem of chaos. The answer to the latter question would require further analysis of the nature of the problem of chaos and of the ability of a principle of the noted structure and claim to respond to it. Fortunately, for the purpose of explaining the role of the principle of nature's purposiveness in addressing the problem of the necessity of empirical laws, an understanding of the principle's structure and claim suffices.

Regulativity

The first thing to note about the principle of nature's purposiveness in the *Critique of the Power of Judgment* is that, despite its purported transcendental necessity, and in contradistinction to the transcendental principles of the *Critique of Pure Reason*, it is not an objective and, in particular, not a constitutive principle of the understanding. Normatively (in terms of its bindingness), Kant presents the principle of nature's purposiveness as a transcendentally necessary principle. Logically (in terms of its structure and claim), Kant presents the principle of nature's purposiveness as "neither a concept of nature nor a concept of freedom because it attributes nothing at all to the object (nature)" (*CJ* 5:184). Instead, Kant insists that the principle of nature's purposiveness is "a subjective principle (a maxim) of the power of judgment" (ibid.). Neither a theoretical principle of object-determination nor a principle of free, rational self-determination, the principle of nature's purposiveness is in fact a technical-practical principle of judgment-determination. As Kant now puts it, the principle of purposiveness is not a constitutive principle of the determining power of judgment, but a "regulative" (ibid.) principle of the reflecting power of judgment. The reflecting power of judgment then just is the faculty of judgment-determination (as opposed to object-determination), and its maxims, including its own transcendental principle, are regulative in the sense of "judgment-governing."

The principle's first distinctive characteristic – its logical character as a subjective, namely, judgment-determining or regulative maxim – raises obvious questions not only about its actual transcendental warrant, but, more fundamentally, even about its candidacy for transcendental status. After all, in Kant's critical philosophy a principle's transcendental status appears to be intimately bound up with its constitutivity or its a priori object-determination. In light of this, a *non*constitutive transcendental principle would appear to be a contradiction in terms. Kant's consideration of regulative principles as candidates for transcendental status – a line of thought Kant initiates in the Appendix to the Dialectic of the *Critique of Pure Reason* – has accordingly led some commentators to conclude that Kant must on occasion be using the term "transcendental" in a nonstandard sense.[10]

It is important to emphasize that there is, in fact, no more than an appearance of a problem here. Kant's official definition of the term "transcendental" carefully avoids narrowing the scope of transcendental cognition to a priori object-determination: "I call all cognition *transcendental* that is occupied not so much with objects but rather with our mode of cognition of objects in general, insofar as this [author's note: this mode of cognition] is to be possible *a priori*" (A11–12/B25). In other words, transcendental cognition is a form of a priori metacognition, "occupied with" – but not necessarily itself in the business of – a priori object-determination. Transcendental cognition is, strictly speaking, a priori cognition of a priori cognition. This type of cognition may (and in the *Critique of Pure Reason* prominently does) include a priori concepts and principles that are "occupied with" our mode of a priori object-determination in no more elaborate a way than by themselves being instances of a priori object-determination (hence, by being determinations constitutive of this or that feature of objects of experience in general). Yet, structurally speaking, such concepts and principles are at best a limiting case. A more natural way in which one type of cognition (transcendental cognition) may be "occupied with" another type of cognition (a priori object-determination) is by reflecting on, determining, or governing it. On purely structural grounds, then, a priori regulative principles of the reflecting power of judgment that in some (yet to be determined) sense are concerned with a priori object-determination actually fit more naturally with Kant's conception of transcendental cognition than even the categories do.

Heautonomy

The candidacy for transcendental status of the principle of purposiveness is thus not threatened by its regulativity or by its nature as a judgment-

[10] Guyer 1990: 28; Kemp-Smith 1992: 552; Grier 1997: 20.

determining (as opposed to object-determining) principle. But the principle of nature's purposiveness is a judgment-determining principle of a peculiar kind. As Kant puts this second main characteristic, the principle of nature's purposiveness "represents the unique way in which we must proceed *in reflection* on the objects of nature" (*CJ* 5:184, emphasis added). The principle of nature's purposiveness is thus a principle that governs the *reflecting power of judgment* itself in its own indigenous transcendental cognitions. It is then not a judgment-determining principle *simpliciter* but, more properly, a judgment-determining principle that determines reflecting judgment (a *reflecting*-judgment-determining principle, if you will, as opposed to a *determining*-judgment-determining principle).

This second characteristic is confirmed by Kant's insistence that the transcendental principle of nature's purposiveness is a principle that "the [reflecting] power of judgment must thus assume . . . as an *a priori* principle *for its own use*" (*CJ* 5:183, emphasis added). Qua judgment-determining or regulative principle (first characteristic), the principle of nature's purposiveness is a principle *of* the reflecting power of judgment. Qua *reflecting*-judgment-determining principle (second characteristic), the principle of nature's purposiveness is a principle *for* (namely, *governing*) the reflecting power of judgment. The particular type of regulative governance of the transcendental principle of nature's purposiveness can then only be described as a case of the reflecting power of judgment's *self*-governance or, as Kant calls it, its "heautonomy": "[t]he [reflecting] power of judgment thus also contains an *a priori* principle for the possibility of nature, but only in a subjective regard, through which it prescribes a law, not to nature (as autonomy), but to itself, for reflection on the latter, (as heautonomy)" (*CJ* 5:185).

Purposiveness

At this juncture one may well ask: if the principle of purposiveness is a principle of the reflecting power of judgment's self-governance and, so, "attributes nothing at all to . . . nature" (*CJ* 5:184), in what sense can it still be considered a principle of *nature's* purposiveness? The third main characteristic of the principle of nature's purposiveness helps answer that question. In prescribing the law to itself for reflection on nature, the reflecting power of judgment demands of itself that it make an *assumption* about nature. Specifically, the principle of nature's purposiveness is the reflecting power of judgment's self-governing (i.e., heautonomous) *demand* for a reflecting (i.e., regulative or judgment-determining) *assumption* that nature exhibits *systematic unity*. As Kant puts the point, the principle of purposiveness prescribes "the unique way in which we must proceed in reflection on the objects of nature" (*CJ* 5:184). This "unique way" is "to *assume* that there is [systematic] unity" (ibid., emphasis added) in nature.

It is important to realize that, even as the *demand* in question is one that can be issued only by the reflecting power of judgment and concerns an *assumption* that must, moreover, be made by that same power of judgment (in short, even as it is a demand *by* and *for* the reflecting power of judgment, or a heautonomous demand), the assumption so demanded does not similarly govern reflecting judgment. The assumption demanded is a judgment-determining assumption that determines *determining* judgment. In this way it retains objective import, if at a remove. Kant says, "The power of judgment's concept of a purposiveness of nature still belongs among the concepts of nature, but only as a *regulative principle* of the faculty of cognition" (*CJ* 5:197, emphasis added). When fully analyzed, then, the transcendental principle of nature's purposiveness emerges as an, in effect, *doubly* regulative principle. It is a *reflecting*-judgment-determining (hence, regulative; specifically, heautonomous) *demand* for a *determining*-judgment-determining (hence, regulative) *assumption* of nature's purposiveness.

Just how a regulative assumption of nature's purposiveness governs determining judgment – and, indeed, governs it with full-fledged transcendental necessity, warding off some form of thoroughgoing cognitive chaos that would otherwise thwart all human attempts at object-cognition – is a question that leads beyond the confines of the structural concerns of the present section and the research aims of the present chapter. I can here suggest only an interpretation.[11] Kant's basic idea appears to be simple. It is that an assumption of the conceptualizability of the world is inscribed into the basic relational (i.e., multiplace) structure of our determining syntheses. According to Kant, in our spontaneous synthesizing activity, we "run through and then take together" (A99) the sensible manifold. Logically speaking, the structure of these syntheses consists in a "running from __ through __ to __" and in a "taking __ together with __" in light of some tertium quid. An important presupposition of our spontaneous syntheses is, accordingly, that discrete run-throughable, reidentifiable elements capable of being taken together are indeed there to enter into them, hence, that nature in fact exhibits a structure we can grasp. This is a presupposition of nature's *purposiveness*, since, for Kant, the only way we can think of

[11] For further discussion, see Teufel 2012: 313–322.

a thing as actually exhibiting graspable structure is to consider it as the effect or product of an associated structuring concept – and Kant's technical term for that type of causality is "purposiveness."[12] It is, moreover, a presupposition implicitly governing *any* determining synthesis without, therefore, itself being a determining synthesis. Accordingly, it is not an object-determining or constitutive principle of the understanding but a judgment-determining or regulative assumption of the reflecting power of judgment – albeit a transcendentally necessary regulative assumption, since, without it, there could be no determining syntheses, hence, no cognition (empirical or otherwise).

6.4 The *Claim* of the Principle of Nature's Purposiveness: Minimalism versus Maximalism

This analysis of the structure of the principle of nature's purposiveness as a doubly regulative principle, heautonomously demanding a transcendentally necessary regulative assumption governing determining judgment, leaves two major questions unanswered. First, as noted, the actual transcendental warrant of the assumption of nature's purposiveness rides on the precise nature of the problem of chaos that this principle of nature's purposiveness is supposed to address. This cannot be pursued further at the present point. Second, and more pertinent to my purposes here, the claim of the principle – the nature of the purposiveness assumed – has not yet been spelled out sufficiently.

Kant's conception of nature's systematic unity is subject to a profound if little-noticed change between the Appendix to the Transcendental Dialectic in the *Critique of Pure Reason* and the Introduction to the *Critique of the Power of Judgment*. In simplest terms, it is a change from a maximalist conception of nature's systematicity to a minimalist conception.

This change is the result of a change in the cognitive faculty in terms of which Kant frames his discussion of systematicity.[13] In the *Critique of Pure Reason*, heuristic maxims of nature's systematic unity were to be grounded in corresponding ideas of reason, which were to have genuine ontological import.[14] Given reason's inherent maximizing tendency, nature's systematicity was, more specifically, conceived in terms of three principles of reason: (1) *unity*, the maximal unity of all natural phenomena as expressions of ever simpler underlying natural forces and, ultimately, of a single

[12] "[T]he causality of a *concept* with respect to its *object* is purposiveness" (*CJ* 5:220).
[13] See Guyer 1990: 17ff. [14] See A660f./B688f.

natural force; (2) *manifoldness*, nature's maximal specificity as reflected in the possibility of ever finer-grained distinctions among natural phenomena; and (3) *affinity*, nature's maximal continuity in the transition between these forms, for example, "*in mundo non datur saltus*" (A228–229/B281).[15] These ontological principles were then reflected in heuristic maxims of scientific inquiry, which exhort researchers to capture that maximal unity, manifoldness, and affinity in their theory construction.

In the *Critique of the Power of Judgment*, by contrast, nature's systematic unity is no longer the object of a set of maximizing ideas of reason. The "principle of nature's purposiveness" is the transcendental principle governing the reflecting power of judgment, not reason. Its claim is that this reflecting power of judgment must (of transcendental necessity) assume that there is "lawful unity" (*CJ* 5:183) in nature. As Kant explains in his helpful elaboration in the paragraph following the "deduction of the present concept [of purposiveness]" (*CJ* 5:184), this amounts to the demand that reflecting judgment assumes that "a cognizable order of nature is possible" (*CJ* 5:183). The shift in associated cognitive faculty from reason to judgment, in combination with Kant's focus on cogniz*ability* suggests that the necessarily presupposed measure of cognizable order cannot consist in nature's maximal unity (lest the bar for cognizability be set impossibly high). The principle's judgment-determining injunction is, instead, that the reflecting power of judgment must (of transcendental necessity) construe nature as if it met a baseline requirement for human cognition. Given the judgment-based (namely, synthesizing) nature of human cognition, this requirement specifically concerns nature's exhibition of at least a minimum of subdivisible, unifiable order. Nature may, naturally, exhibit order much beyond minimal cognizability. Still, the principle of purposiveness pointedly concerns nature's satisfaction of that *threshold* of cognizability – wherever it may lie, and whether nature exceeds it or not. Accordingly, by comparison with reason's ideas of nature's maximal unity, the regulative principle that judgment assume that nature meets requirements of cognizability is minimalist in ambition.[16]

[15] See A662/B690.

[16] Kant's declaration at the end of §76 that the "concept of the purposiveness of nature [is] a subjective principle of *reason* for the power of judgment" (*CJ* 5:404, emphasis added) appears to challenge this reading. It also appears to contradict the heautonomy of the reflecting power of judgment. However, at that point in the Dialectic of the Teleological Power of Judgment, Kant is talking not about the transcendental principle of nature's purposiveness but about a global teleological maxim for research for which reason seeks authorization by extending the apparent necessity of judging organisms teleologically.

6.5 Methodological Maxims

What kinds of methodological advice might such a minimalist principle of nature's systematic unity ground? The short answer is: minimalist maxims. At first, however, this does not seem to be how Kant proceeds. As we saw earlier, in the lead-up to the transcendental deduction of the principle of nature's purposiveness in §V of the Introduction, Kant says that nature's purposiveness "shines through" in a set of "maxims of metaphysical wisdom" (*CJ* 5:182). The research-governing maxims Kant mentions under this heading are:

> "Nature takes the shortest way" (*lex parsimoniae*); "it makes no leaps, either in the sequence of its changes or in the juxtaposition of specifically different forms" (*lex continui in natura*); "the great multiplicity of its empirical laws is nevertheless unity under a few principles" (*principia praeter necessitatem non sunt multiplicanda*); and so on. (*CJ* 5:182)

These *pre*deduction methodological maxims are essentially equivalent to and clearly intended as a reminder of the subjective, maximalist heuristic advice discussed in the Appendix to the Dialectic of the *Critique of Pure Reason*.[17] But the suggested continuity between this maximalist heuristic advice and the minimalist principle of nature's systematic unity Kant is about to offer is in fact more rhetorical than substantive. After all, Kant's attempt to establish the *transcendental necessity* of objective maximalist ideas of nature's systematic unity in the Appendix to the Dialectic proves unsuccessful (issuing in the effort to deduce a subjective minimalist principle of nature's systematic unity in the *Critique of the Power of Judgment*). By contrast, Kant's attempt, in the Appendix, to ground maximalist *heuristic* advice in precisely those objective maximalist (albeit nonnecessary) ideas of reason arguably succeeds.[18] Kant's "maxims of metaphysical wisdom" (*CJ* 5:182) need no further grounding than the one provided in the Appendix to the Dialectic, and Kant is not here trying to reassign them to the reflecting power of judgment.

The distinction between, on the one hand, methodological advice rooted in ideas of nature's maximal unity and, on the other hand, methodological advice rooted in a principle of nature's minimal cognizability is on full display when one compares the *pre*deduction "pronouncements of metaphysical wisdom" that supposedly "shine... through" (*CJ* 5:182) in a principle of nature's purposiveness with the set of *post*deduction

[17] See A658/B686. [18] See Grier 1997: 14f.

methodological maxims that actually "express" (*CJ* 5:185) the principle Kant deduces. The latter methodological maxims comprise:

> the following propositions: that there is in nature a subordination of genera and species that we can grasp; that the latter in turn converge in accordance with a common principle, so that a transition from one to the other and thereby to a higher genus is possible; that since it seems initially unavoidable for our understanding to have to assume as many different kinds of causality as there are specific differences of natural effects they may nevertheless stand under a small number of principles with the discovery of which we have to occupy ourselves, etc. (*CJ* 5:185)

In this list of *post*deduction "propositions," the confident maximalist claims of Kant's *pre*deduction "pronouncements" – that is, that nature takes the "*shortest* way," makes "*no* leaps," and that the multiplicity of empirical laws (at bottom) "*is* unity" (*CJ* 5:182, emphasis added) – are conspicuously absent. They have been replaced by sober references to the possibility of added layers of organization and the belief that, at any given level of organization, higher genera remain *possible*, that a multiplicity of phenomena *may* stand under a small number of principles, etc.

Methodological advice to judge nature as unified rooted in a principle of nature's minimal cognizability thus consists in a modest call to explore whether nature surpasses the level of order required for cognizability whose postulation the principle of nature's purposiveness demands. Such advice is expressed by subjectively necessary maxims advocating (based on nature's presumed subdivisibility and unifiability) that – for any given level of order discovered – there may yet be a higher genus, a lower species, or further continuity discoverable. It is advice concerned with expanding the encountered extent of nature's lawful unity, as it were, from the inside: a level up, a level down, a level over. Crucially, this type of advice is not concerned with pursuing nature's lawful unity based on an extraneous rational commitment to nature's thoroughgoing systematicity.

6.6 The "Maxim of the Lawfulness of Empirical Laws"

What bearing, then, do this transcendental principle of nature's threshold-cognizability and its associated methodological maxims of scientific inquiry have on the question of the necessity of empirical laws of nature?

Kant begins his discussion of the necessity of empirical laws of nature in §V of the Introduction in functional terms: we cognize "the lawful unity in a connection" *as* lawful unity, namely, as noncontingent and necessary, "in accordance with a necessary end (a need) of the understanding"

(*CJ* 5:184). The cognition of candidate empirical causal generalizations as lawful and necessary thus serves as a cognitive *means* to some further cognitive *end*.

It is odd, however, that Kant should characterize cognizing the lawfulness of empirical causal generalizations, or "each of these ways [of being a cause]" (*CJ* 5:184), as cognitively instrumental. The lawfulness of the connections in question follows analytically from their nature as *causal* generalizations, by the principles laid down in the Analytic of Principles. And the necessity of such lawful unity is, in turn, analytically contained in the notion of a law. The understanding thus adopts the means in question, that is, cognizing causal generalizations as lawful and necessary, by cognitive default, which is to say, *not* in pursuit of some further cognitive end (necessary or otherwise). As Kant explains this automatism, each of the ways in which a product of nature can be a cause "ha[s] its rule, which is a law, and hence brings necessity with it" (*CJ* 5:183). At best one might perhaps say that the means is here identical to the end or that the means in question is an act of *cognizing* while the end in question is the corresponding state of *cognition*.

Analytically inescapable as cognizing empirical causal generalizations as lawful and necessary may thus be for the understanding, perhaps Kant considers it a means to an end also because so cognizing empirical generalizations is nevertheless a somewhat gratuitous perspective the determining power of judgment takes on a subset of suitably formed empirical cognitions. It is a way of *regarding* empirical generalizations that not only is objectively unverifiable but does not affect our ability to *form* these cognitions in the first place (it is, instead, a consequence of their formation). Cognizing empirical causal generalizations as necessary is not among the transcendental conditions of object-cognition. It may be a perspective the understanding takes by default, but it is transcendental-logically dispensable and perhaps in this sense (like a means) adoptable.

Indeed, more than merely gratuitous, cognizing empirical causal generalizations as lawful and necessary is problematic in yet another way, crucial to Kant's exposition: it is prima facie false. Empirical generalizations are known to be known a posteriori and, hence, known to be contingent, *not* necessary. Kant says that we cognize the lawful unity in a connection as lawful and necessary "but yet at the same time as in itself contingent" (*CJ* 5:184). For all we know, then, the understanding's cognition of empirical generalizations as necessary might be little more than a self-congratulatory fiction.

How, then, can an empirically unverifiable, transcendental-logically dispensable default-cognition serve a necessary end of the understanding?

What needs sharpening, here, is the formulation of the understanding's cognitive end itself. The understanding's end is not in fact, as first appeared, the cognition of empirical causal generalizations as necessary. The understanding's end is, instead, to be *entitled* to that default cognition in spite of the generalizations' known epistemic contingency. This is a necessary end in the sense that the understanding must satisfy it or else suffer the inconsistency of wittingly and inescapably regarding its most accomplished empirical cognitions in patently incongruous ways – as, at the same time, contingent and noncontingent. It is thus an end whose satisfaction the understanding strongly desires (on pain of inconsistency), though not one whose satisfaction the understanding requires in order to carry out its cognitive business. It is, in short, a *subjective* end of the understanding or, as Kant calls it, "a need" (*CJ* 5:184). To the extent that there is a problem of the necessity of empirical laws in critical philosophy, it concerns the satisfaction of *this* need.

The cognitive means required in order to satisfy the need in question is then, similarly, not the analytical automatism by which we do in fact consider empirical causal generalizations as lawful and necessary. It is, instead, to adopt a perspective on those generalizations that might warrant our default ascription of lawfulness and necessity to them. Now, to consider empirical generalizations as lawful and necessary despite their known epistemic contingency is to consider them lawful and necessary in an objective and yet nonempirical – hence, in a *metaphysical* – sense. It is to consider them as laws expressive of and structuring transcendent reality, or as laws absolutely. From the vantage point of the *Critique of Pure Reason*, this would seem to be a cognitive stance unavailable as an even remotely legitimate perspective on empirical causal generalizations or candidate empirical causal laws. The role of the transcendental principle of nature's purposiveness in the *Critique of the Power of Judgment* (and of its associated maxims of scientific inquiry) is to underwrite that cognitive stance concerning empirical causal laws – and, hence, to support the cognitive means we *must* adopt if we are to be entitled to our default ascription of necessity to those laws – as a stance we in fact *may* adopt.

In short, the transcendental principle of nature's purposiveness serves as the *rationale* backing up an instrumental cognitive maxim of the following sort: "if you want to be *entitled* to consider empirical causal laws as lawful and necessary despite their known epistemic contingency" – as the understanding wants to be, given the "need" mentioned – "then you must presume them to be metaphysically necessary, namely, to be reflective of an underlying order of nature." The epistemic consequence of a counterfactually presumed absence of the transcendental principle of purposiveness with which we started would then be the failure to be entitled to the presumption spelled out in the maxim's consequent and, a fortiori, the failure to be *entitled* to consider empirical causal laws as lawful and necessary, as we nevertheless do by cognitive default. Absent a principle of nature's purposiveness, our "need" could not be satisfied and the Humean critic of the assumption of the necessity of empirical causal laws would, accordingly, be vindicated.

6.7 Conclusion

It is now relatively easy to see how the structure (see Section 6.3, above) and claim (see Section 6.4, above) of the transcendental principle of nature's purposiveness, together with the principle's associated methodological maxims of scientific inquiry (see Section 6.5, above), serve to support this "Maxim of the Lawfulness of Empirical Laws" (see Section 6.6, above).

To say that the principle of nature's purposiveness is a heautonomous transcendental principle of the reflecting power of judgment is to say that the possibility of object-determining syntheses (and, a fortiori, the possibility of human cognition as such) necessarily presupposes that nature exhibits at least a threshold of cognizable order. All actual human cognition – simply by being self-conscious awareness of causally interacting substances in space and time – then in turn confirms the existence of (at the very least) that heautonomously presumed minimum of synthetically unifiable and analytically subdivisible order. Moreover, by methodically following the principle's associated heuristic maxims, urging us to look for further levels of unifiability and subdivisibility, we are justified in (and, so, entitled to) the assumption that, whenever we come across further "lawful unity in a combination" (*CJ* 5:184), that lawful unity is continuous with the original (justifiably presumed and cognitively confirmed) minimal order. Hence, we are entitled to the assumption that empirical laws formulated in adherence to these maxims reflect the way the world is and, so, are laws expressive of and structuring transcendent reality, or laws absolutely. If this is a perspective on empirical causal laws that we, accordingly, *may* adopt, then the Maxim of the Lawfulness of Empirical Laws says that it is, moreover, a perspective we *must* adopt, *if* we want to be entitled to our cognitive default of considering empirical causal laws lawful and necessary despite their known epistemic contingency.

Still, if following the Maxim of the Lawfulness of Empirical Laws, as backed by the transcendental principle of nature's purposiveness and its associated methodological maxims, provides no more than entitlement to the mere assumption of the lawfulness and necessity of empirical causal laws, not much seems to have been gained over the regularity theorist who does quite well without such additional presumptions of causal realism. This is the point at which the philosophical role of the transcendental-philosophically justified methodological maxims comes into fuller view. While the picture of the natural world that emerges from our scientific theorizing in principle remains open to revision (all the way from fine-tuning corrections to large-scale paradigm shifts), the iterative implementation of these methodological maxims offers transcendental-philosophical warrant for the belief that the ever finer-grained, more expansive, and more integrated picture of the natural world that emerges is not simply increasingly coherent but tracks metaphysical truth asymptotically. This much, though not more, is what Kant's heautonomous transcendental principle of nature's purposiveness in the *Critique of the Power of Judgment* has to offer in response to the Humean skeptic.

PART III

Nomic Necessity and the Metaphysics of Nature

CHAPTER 7

Kant's Necessitation Account of Laws and the Nature of Natures

James Messina

7.1 Introduction

Kant's Critical works contain a revolutionary account of the epistemology and metaphysics of the laws of nature. In terms of the epistemology, Kant claims that we can have synthetic a priori knowledge of some very general laws of nature, a class that includes the so-called transcendental laws of nature, such as the law that every alteration has a cause, along with the closely related laws of mechanics, such as the law that "every change in matter has an external cause" (*MF* 4:543). In terms of the metaphysics, Kant claims that these laws do not merely describe the regularities that happen to obtain in our world; instead, they are governing principles endowed with nomic necessity. On these epistemological and metaphysical points, the Critical Kant's view contrasts sharply with Hume's. For Hume, to believe that such-and-such is a law is to believe, on the basis of experience and custom, that there is a certain constant conjunction among events in our universe (Bs follow As). Laws have no governing force – they do not, as it were, force events in the world to fall into any particular pattern – nor do they rest on any underlying necessary connections in nature.

One of Kant's most radical claims arises in the course of his attempt to explain how the most general laws are necessary, governing, and knowable a priori: the pure understanding is their legislator. Call this the Legislation Thesis. As Kant writes in the *Prolegomena*: "even though it sounds strange at first, it is nonetheless certain, if I say with respect to the universal laws of nature: *the understanding does not draw its (a priori) laws from nature, but prescribes them to it*" (*Prol* 4:320).[1] The reason objects of experience must obey the transcendental laws, and the reason we can know them a priori, is

[1] Cf. A127–128, B165, and *CJ* 5:167.

that they are the understanding's own subjective conditions of the possibility of experience.

These features of Kant's account of laws, painted in very broad strokes, are generally agreed on. There is, however, considerably less agreement when it comes to Kant's views on another class of laws: the particular, empirical laws of nature, which figure into specific sciences, like chemistry. What is it to be a *particular law* for Kant – what does their lawfulness consist in?[2] Do such laws have the governing character and nomic necessity of transcendental laws? In what manner are the particular laws epistemically and/or metaphysically grounded in the transcendental laws? How, if at all, are such laws knowable?

Up until recently, there were two major interpretations of Kant's account of particular laws of nature. According to Michael Friedman's interpretation, by applying our knowledge of the synthetic a priori laws to the relevant empirical data we are able to derive knowledge of the particular empirical laws. Particular laws themselves inherit the necessity and prescriptive character of the laws of nature that provide their epistemic basis. By contrast, on the Best System Interpretation, defended by Philip Kitcher, among others, there are certain empirical generalizations that we are justified in taking to be necessary and thus to be laws, but only in the context of an ideal systematization of the empirical data (Kitcher 1986).[3] Metaphysically, particular laws are, or correspond to, the statements that would be regarded as laws in such a system. In apparent contrast to these older readings, the so-called Necessitation Account[4] holds that particular laws have a necessity that is "as it were built right into the nature of things" (Watkins 2005: 346). Empirical objects that share a certain nature (e.g., salt) are bound to behave in accordance with certain rules given those natures. The Necessitation Account suggests that Kant has a "bottom-up"[5] metaphysics of particular laws, whereby laws are grounded in the nature of things as opposed to being superimposed on them. As for the epistemology of laws, the Necessitation Account denies that particular laws in most cases are knowable by us at all. On both of these points, concerning the

[2] I owe this formulation of this question to Kreines 2009.

[3] I borrow this label for Kitcher's position from Kreines 2009. The aptness of the label is supported by remarks like the following: "For Kant, the laws of nature are just the generalizations that would figure in the best unifying system in the limit of rational inquiry" (Kitcher 1996: 412).

[4] Again, I borrow this label from Kreines 2009.

[5] I borrow the language of top-down and bottom-up – which I explain below – from Ott 2009: 5–6 and Massimi 2014: 492–493.

metaphysics and the epistemology of laws, the Necessitation Account challenges the received picture of Kant.[6]

In this chapter, I have several aims. First, I will extend the case for the Necessitation Account, adding to the evidence that has been adduced by its proponents. Second, I will argue that the evidence supports an *expanded version* of the metaphysics of the Necessitation Account, whereby *all* laws of nature are necessary rules that are immanent to certain natures. Third, I will examine Kant's ontology of natures in order to clarify the metaphysical relationship between natures and laws, using as a foil for Kant's view a contemporary "bottom-up" version of the Necessitation Account. Here I argue that, while the existence of particular laws of nature is tied to the existence of empirical natures, they also have a priori grounds of possibility (not all of which are knowable by us) from which their modal force and content come. One interesting result here is that Kant's account is immune to the charge that has been leveled against some contemporary bottom-up versions of the Necessitation Account, namely, that they render laws of nature otiose (Mumford 2004: 121). Fourth, I respond to some possible objections to my expanded Necessitation Account.

7.2 Three Interpretations of Kant on Particular Laws of Nature

In contrast to the transcendental laws of nature, which are "pure" and govern the behavior of all objects of experience without exception, all more specific laws of nature have an experiential element and are restricted to a particular domain of objects. As a result of their empirical character, Kant says that "particular laws ... *cannot* be *completely derived* from the categories, although they all stand under them" (B165).[7] Particular laws do not admit of either a metaphysical or a transcendental deduction, nor can they be justified through construction in pure intuition. Prior to all experience, it is impossible for the human understanding to know what empirical laws obtain and whether and to what extent those laws are related to each other in a tidy, hierarchical way (*CJ* 5:183).

Kant's various commitments regarding empirical laws have put close readers into a quandary.[8] As we have just seen, Kant claims that (1) we don't have a priori knowledge of empirical laws. Kant also seems to claim

[6] It is noteworthy in this regard that Brian Ellis places Kant alongside Descartes and Newton as a proponent of the top-down view (Ellis 2001: 1).

[7] Cf. A127–128. [8] My formulation here has been influenced by Kitcher 1986: 208.

that (2) empirical laws, like all laws, involve some sort of necessity;[9] (3) knowledge of necessity cannot be obtained empirically;[10] and (4) we have knowledge of empirical laws.[11] It is not immediately obvious how (1)–(4) are to be reconciled with one another, particularly if one assumes that all knowledge is either empirical or a priori. In the effort to reconcile Kant's various commitments regarding particular laws, various interpretations have been developed.

Michael Friedman attempts to resolve the apparent tension in Kant's commitments by claiming that, while particular laws have an empirical component, this does not preclude an priori grounding as well – a grounding in a priori knowable laws (Friedman 1992b: 181). Textual evidence for such a view can be found in passages like the following:

> Even laws of nature, if they are considered as principles of the empirical use of the understanding, at the same time carry with them an expression of necessity, thus at least the presumption of determination by grounds that are a priori and valid prior to all experience. But without exception all laws of nature stand under higher principles of the understanding, as they only apply the latter to particular cases of appearance. (A159/B198)

> There are therefore certain laws, and indeed a priori, which first make a nature possible; the empirical laws can only obtain and be found by means of experience, and indeed in accord with its original laws, in accordance with which experience itself first becomes possible. (A216/B263)

Friedman understands Kant to be claiming here that particular empirical laws, while not derivable from the a priori laws of nature in isolation from experience, can be derived from such laws when they are applied to the relevant empirical content (Friedman 1992b: 174).

There are two classes of particular laws to consider. On the one hand, there are the laws of mechanics. These are deduced by applying the transcendental laws to the empirical concept of matter (Friedman 1992b: 185). On the other hand, there are "mixed" laws like the law of universal gravitation. Knowledge of such laws requires a substantial empirical component. For example, in the case of the law of universal gravitation, it depends on Kepler's laws, which are strictly speaking inductively obtained rules, not laws, because they lack necessity (*Refl.* 18:176 [R5414]). While induction on its own cannot give us knowledge of the law of universal gravitation, we can deduce this law from the inductively attained

[9] For the claim that laws involve necessity, see A113, *MF* 4:469, A159/B198 (quoted below), *CJ* 5:184–185, and *Refl* 18:176 (R5414).

[10] See, e.g., A1 and B3. [11] E.g. *CJ* 5:184–185, quoted in Kitcher 1986: 208.

regularities by availing ourselves of the a priori laws of mechanics, along with certain assumptions (Friedman 1992b: 175–178).[12]

Friedman's account, which we might call the Derivation Account (DA), does not merely concern the epistemology of particular laws. It also has implications for the metaphysics of such laws. On the DA, particular laws are those the knowledge of which can be attained by applying a priori principles to experience (Kreines 2009: 528). Particular laws possess genuine nomic necessity, one that is "injected" into them by the transcendental laws (Friedman 1992b: 175). They are not, as on the Humean account, mere regularities that happen to obtain in nature, but rather are rules that govern how nature must behave.

The DA contrasts with the Best System Interpretation (BSI) defended by, among others, Philip Kitcher.[13] Kitcher's way of resolving the tension among (1)–(4) is to claim that

> taken individually, statements that we normally count as laws can only be regarded as empirical and contingent. But, we are required to systematize the body of our beliefs, and as consequence of the systematization, some statements (in fact those we count as laws) come to be credited with necessity. (Kitcher 1986: 209)

Considered apart from other beliefs, a belief in a true empirical generalization that is arrived at through induction does not constitute knowledge of a law, since induction cannot justify a claim to necessity. This is how Kitcher construes Kant's commitment to (3) above. But the belief that the generalization has necessity and is a law is justified *if* the generalization plays a particular role[14] in an ideal systematization of our empirical beliefs (Kitcher 1986: 210). Such an ideal systematization would be based on all empirical data and be guided by certain methodological rules. This is how Kitcher makes sense of (4). Kant's accounts of systematic unity, and his views on the methodological rules that guide the process of systematization, can be found in the Appendix to the Transcendental Dialectic and in the introduction to the *Critique of Judgment*. These methodological rules tell us, among other things, to bring particular empirical beliefs into a

[12] While Kant's remarks on the law of universal gravitation at *Prol* 4:321 might seem to suggest that Kant believes that it is possible to give a purely geometric derivation of the law, it is generally agreed that this is not the right way to read the passage.

[13] Other versions occur in Buchdahl 1965, Brittan 1978, and Allison 1996.

[14] Unfortunately, it is unclear exactly what role the generalization would have to play in order to be justifiably deemed a law. Presumably, the generalization would have to be essential for unifying lower-level regularities and be couched in terms of empirical concepts that similarly unify lower-level empirical concepts.

hierarchy of progressively more general beliefs that has the right balance of simplicity, fruitfulness, and continuity. The belief hierarchy thereby constructed will contain beliefs in empirical generalizations that we are justified in regarding as necessary laws, given the role they play in the construction.

The BSI is not simply an interpretation of Kant's epistemology of particular laws. Like the DA, it is also an interpretation of the metaphysics. For Kitcher, particular laws of nature will be, or will correspond to, those empirical generalizations that are deemed to be laws in a best system (Kitcher 1986: 214). In contrast to the DA, the necessity of particular laws is not "injected" into them by the transcendental laws, but has a different provenance in the cognitive faculties responsible for systematization: reason and/or reflective judgment. The necessity is of a kind different from the necessity of transcendental laws and is harder to get a grip on. With regard to the latter point, it is not clear how connecting a generalization systematically with other generalizations could change its modal character. With regard to the former point, the laws in question cannot be *governing* principles. On the BSI,[15] particular laws do not *explain* why their associated regularities must obtain, but rather are part of an ideal codification of the regularities that happen to obtain in our world (Kreines 2009: 549–550). For these reasons, the BSI makes Kant a kind of Humean about particular, empirical laws.[16]

Recently, a new reading of Kantian particular laws of nature has emerged, the so-called Necessitation Account (NA), aspects of which have been defended by Watkins and Kreines. According to Watkins, "laws of nature are nothing other than the laws of the natures of things. That is, the laws of nature that hold in a given world are a function of the natures that are instantiated in that world" (Watkins 2005: 335). Kreines speaks of particular laws obtaining in virtue of the nature of specific kinds of things.

[15] Recently, McNulty has developed an interesting reading of Kant's account of particular laws of nature that builds on some of the ideas of the BSI while avoiding its shortcomings. On McNulty's "ideational interpretation," ideas of pure reason play a crucial role in accounting for necessity (McNulty 2015: 2). Unfortunately, I do not have space to evaluate McNulty's reading here.

[16] This is notwithstanding Kitcher's own claims about the anti-Humean character of Kant's view. Kitcher takes Kant to be arguing in the Second Analogy that "there is no justifying straightforward empirical claims, descriptions of Hume facts, without justification of causal claims." He takes Kant to be arguing in the Appendix to the Transcendental Dialectic that these causal claims are justified "through the incorporation of statements within a unified system" (Kitcher 1986: 221). These causal claims are precisely the kind whose legitimacy was contested by Hume. They "impl[y] generalizations that legislate for unactualized possibilities" (Kitcher 1986: 219). The problem is that it is far from clear that Kitcher's Kant can have such a robust notion of causation and of laws.

He uses the example of the solubility of salt. If indeed it is a law that salt is soluble, this is so because of the nature of the kind salt (Kreines 2009: 531–532). Anything with this nature will necessarily be disposed to dissolve in water. Like the DA, the NA takes particular laws to possess necessity and governing force. They do not merely reflect regularities that happen to obtain, but rather also serve to explain them. However, in apparent contrast to the DA, the necessity in question "is built right into the nature of things" (Watkins 2005: 346).

The NA suggests that Kant is working with a "bottom-up" model of laws. While top-down models take the laws of nature to be transcendent principles that are imposed from the outside on things that could retain their essential properties apart from those laws, bottom-up necessitation accounts deny these claims: laws supervene on the natures and/or essential properties of objects (Ellis 2001: 1; Ott 2009: 5–7). Historically, top-down models of laws invoked God as a legislator, identifying laws of nature with features of God's will, such as divine general volitions on the Occasionalist position. Contemporary top-down models of laws invoke universals that stand in sui generis necessitating relations to each other, relations that vary from possible world to possible world, resulting in corresponding changes to the laws.[17] Such a view disconnects laws from the essential properties and natures of the things that they govern. By contrast, so-called dispositional essentialism, defended by Brian Ellis, is a contemporary bottom-up model of laws that says that laws of nature derive from the essential properties of things and accompany those properties in all possible worlds (Ellis 2001: 4).[18]

That should suffice for now as a preliminary characterization of the metaphysics of laws ascribed to Kant on the NA. What about the epistemology? Kreines's manner of resolving the apparent conflict between (1)–(4) is to deny (4): the particular laws of nature are not, except in the very special case of the laws of mechanics, knowable.[19] In contrast to

[17] E.g. Armstrong 1983, Dretske 1977, and Tooley 1977. For a helpful summary, see Mumford 2004. Admittedly, Watkins initially likens Kant's account of laws to Armstrong's top-down model. But in his discussion, he stresses weaknesses in Armstrong's account that derive from (what I am calling) its top-down aspects and corresponding ways in which Kant's account avoids these problems, though Watkins appears officially noncommittal about whether Kant's account is bottom-up (Watkins 2005: 402–407). At one point, Kreines professes himself agnostic on the question about whether particular natures could retain their identity while obeying different laws (Kreines 2009: 533). But this seems to run contrary to the idea that the natures of kinds provide a deep explanation for why the regularities that obtain do and must obtain (since it would be possible for those natures to exist without supporting those regularities); moreover, it is hard to see what the nature of something like salt could consist in if it is not bound up with obeying certain laws.

[18] Ellis 2001: 4.

[19] It is unclear whether Watkins agrees with Kreines on this point.

Friedman and Kitcher, who define what particular laws are in terms of the ways in which we come to know them as laws, Kreines separates the metaphysical from the epistemic question. Once we do that, there is no contradiction in the idea that there are nomically necessary, governing laws, which we are nevertheless prevented from knowing as laws because of our epistemic limitations (Kreines 2009: 528). The limitations are due to the fact that particular laws involve empirical kinds whose properties and relationships to each other we can generally find out only through empirical intuition. And empirical intuition can tell us only what patterns happen to be the case, not that must be the case (Kreines 2009: 540). The only case where we can have knowledge of particular empirical laws is when the laws involve a single kind that stands in a particularly close connection with our pure intuition of space: namely, matter. This is true of the laws of mechanics, but it is a very special case (Kreines 2009: 540, 543).[20]

7.3 A Case for an Expanded Version of the NA

In this section, I evaluate the case for the metaphysics of the NA. After first adding to the evidence provided by Watkins and Kreines, I then raise some clarificatory questions and objections that I will address in subsequent sections.

Watkins's case for the metaphysics of the NA rests on a sophisticated, historically sensitive interpretation of Kant's model of causality. Crucial to this model is the notion of causal powers that substances exercise in accordance with their natures. Watkins also provides an attractive reading of the resolution to the Third Antinomy wherein the idea that laws of nature are laws of natures plays a crucial role (Watkins 2005: 334).[21] Kreines's case for the metaphysics of the NA begins with some remarks in the Transcendental Dialectic that he takes to commit Kant to a

[20] One might wonder what implications Kreines's view has for our knowledge of particular causal relations. If he were to construe Kant as holding that knowledge of a particular causal relationship requires knowing some particular empirical kind along with the laws sustained by that kind, then Kreines would have to say that in general Kant precludes our knowing particular causal relations. However, Kreines does not commit himself to the antecedent here.

[21] The idea, briefly, is that my phenomenal self's being determined by prior events and the laws of nature to, say, ignore the pleas of the homeless person on the street is compatible with this being a free act because my noumenal self freely decided the nature of my phenomenal self, and the latter in turn partly grounds the particular laws of nature my phenomenal self is subject to. In this way, my noumenal self's choice helps to determine the laws of nature, which includes laws of the nature of my phenomenal self.

particular view of explanation, which he calls the *simple intuition*: "explanation must provide information about an underlying condition on which an explanandum really depends" (Kreines 2009: 531). Now, we typically invoke laws to explain regularities, like that Bs regularly follow As. Kant's view of explanation rules out a Humean understanding of particular laws because then laws will merely restate the regularity in question. Instead, the simple intuition points us toward a model of nomic explanation whereby laws "provide information about an underlying condition on which the regularity *depends*, namely, the nature of the kind" (Kreines 2009: 532). Laws based in natures will explain not just why the Bs we have seen have regularly followed As but also why any future or counterfactual cases of As would have to be followed by Bs. Admittedly, the DA appears similarly well equipped to accommodate the simple intuition. However, Kreines argues that this reading would have the unacceptable consequence of denying the status of laws to any laws that cannot be derived from a priori principles, as appeared to be the case for chemical laws during Kant's time[22] and that will be the case for any laws that involve fundamentally distinct kinds (Kreines 2009: 541).

I think we can add to the case for the metaphysics of the NA. A philosophical consideration that supports the idea that Kant does not in general make the conditions on being a law dependent on the conditions under which laws are known, as on the DA and BSI, is that he countenances at least the possibility of lawfulness at the level of things-in-themselves, despite our lack of epistemic access to them: "The lawfulness of things-in-themselves would necessarily pertain to them even without an understanding that cognizes them" (B164). Only a nonepistemic account of laws, such as the NA, is consistent with the idea of lawfulness at the level of things-in-themselves – particularly if the laws are governing principles endowed with nomic necessity.[23]

The NA also gains some modest support from the details of the development of Kant's thinking about laws. Michela Massimi has argued persuasively that the pre-Critical Kant endorses a dispositional essentialist model of laws, whereby as she puts it, "physical necessity . . . is grounded in nature's capacities and natural powers, from which laws of nature are read off" (Massimi 2014: 493). In texts such as the *Universal Natural*

[22] McNulty 2015: 2 makes the same point.

[23] I am assuming here that Kant would prefer a uniform account of what it is to be a law of nature to one that would make the lawfulness of particular laws fundamentally different from the lawfulness of noumenal laws and transcendental laws.

History and *Only Possible Argument*, Kant claims explicitly that the necessity of the laws of mechanics is rooted in the essence of matter, such that it is impossible for matter to exist as matter and yet operate according to different laws (*OPA* 2:100).[24] The necessity of other laws likewise follows from the essences of the relevant kinds of things: "But that a celestial body in its liquid state should, entirely necessarily and as a result of such universal rules, strive to assume a spherical form . . . that is inherent in the essence of the thing itself" (*OPA* 2:102). For the pre-Critical Kant, the essence or nature of a thing involves fundamental causal powers (or forces) and necessary rules that direct their operations. Particular laws cannot exist without there being causally active natures to support them, and those natures cannot themselves exist without following these particular laws. In this respect, natures are "tied" to laws (*UNH* 1:227).

The character of Kant's pre-Critical account of laws can be brought out by comparing it with some of the historically available models of laws of nature that were likely to have influenced him. On the one hand, there is the top-down model of the laws of nature offered by the Occasionalists. For Malebranche and those of his ilk, laws of nature are simply God's general volitions. The things that exist are not essentially bound to obey the laws that they do – God could in principle have created those same things while having them obey different laws. Finite substances lack causally active natures; it is the laws superimposed on them that are the true causes of their properties, relations, and behaviors. Some Newtonians, like Newton himself and Roger Cotes, are also attracted to a top-down model of laws (Massimi 2014: 494–495; Ott 2009: 7), though they do not go as far as the Occasionalists in denying to finite substances causally efficacious natures. By contrast, Leibniz's account of laws of nature is bottom-up. For Leibniz, regularities in nature are to be explained in terms of the essential natures of substances; these natures are, or include, forces or dispositions that operate according to in-built, necessary rules (Adams 1994: 313–314; Rutherford 1993: 140–141).[25] Any occurrences that are not explicable through the natures of substances are *supernatural*.[26] Insofar as the Occasionalists take all occurrences in the natural world to be this way, and the Newtonians everything involving gravity, Leibniz complains that

[24] This is how I understand Kant's reference to "logical necessity" (*OPA* 2:100).

[25] Leibniz describes natures in *On Nature Itself* as "created, active force" (Leibniz 1989: 156) and "a certain efficacy [that] has been placed in things, a form or force" (Leibniz 1989: 159).

[26] In the *Discourse on Metaphysics* he says that "that which is limited in us could be called our nature or our power and in that sense, that which surpasses the natures of all created substances is supernatural" (Leibniz 1989: 49).

they unduly multiply the supernatural (Leibniz 1989: 143, 336). Of these accounts, the pre-Critical Kant's is clearly closest to the Leibnizian. This is what allows Kant to work with a broadly Leibnizian account of supernatural occurrences in the *OPA*, whereby these involve events that are not explicable through the forces and rules inherent in the natures of things (*OPA* 2:104ff.).

Why not think Kant's pre-Critical account of laws is just another of the fanciful dreams that Kant dreamed during his dogmatic slumber? There are various considerations that point toward Kant's continued adherence to some version of a bottom-up necessitation account in the Critical period, despite radical shifts in his thinking about the natural world and the conditions of our knowledge of it.

First, there is the fact that Kant does not abandon the language of essence and nature in the Critical Period. It continues to occupy a prominent place in his writings on the topic of natural science, for example, the *Metaphysical Foundations* (*MF* 4:467–468). Moreover, there is a natural way of reading this text in which its goal is to reveal as before, albeit within a radically new philosophical framework, the forces and laws that necessarily attach to all matter in virtue of its nature. The causal powers essential to matter include attraction and repulsion, while the laws include Kant's three laws of mechanics and the law of universal gravitation.

Second, there are texts from the late pre-Critical through to the Critical period in which Kant underscores the close relationship between laws and natures. In the Metaphysics L_1, for example, transcripts of lectures given in the 1770s, Kant says that "every nature has laws" (28:216). In a thematically related reflection from the 1780s, he says that something is "contrary to nature [*Naturwiedrig*] insofar as it contradicts [*wiedersteht*] *the laws of the nature of a thing* [*den Gesetzen der Natur eines Dinges*]" (*Refl.* 18:180 [R5432], emphasis added). The context of the former remark is significant insofar as it reveals Kant's continued adherence to a Leibnizian account of the supernatural, an account that fits naturally with a bottom-up model of laws.

Such claims are not confined to unpublished reflections and lecture transcripts. In the introduction to the *Critique of Judgment*, Kant writes that

> specifically different natures, besides that which they have in common as belonging to nature in general, can still be causes in infinitely many ways; and each of these ways must (in accordance with the concept of a cause in general) have its rule, which is a law, and hence brings necessity with it, although given the constitution and the limits of our faculties of cognition we have no insight into this necessity. (*CJ* 5:183)

What such passages indicate is that in the Critical period Kant continues to think that each empirical nature is associated with distinctive causal powers and general, necessary rules according to which these powers operate. Specific empirical natures cannot exist without bringing with them laws, which lead to regularities in the effects produced by them. This is so even if we cannot know what the laws are and whether and to what extent there are any substantive connections between the empirical laws governing different natures.

Significantly, though, the Critical Kant's views on the connection between laws and natures do not appear to be restricted to *particular* empirical natures and *particular* empirical laws. Kant also has a notion of "Nature in general" (referenced in the immediately quoted passage), by which I take him to mean both the essential structural aspects of the empirical world *and* a most general nature common to, and contained in, all particular empirical natures, in the same way that the nature of metal is common to, and contained in, the natures of gold and lead.[27] Just as particular, empirical natures are necessarily accompanied by their corresponding particular laws, so Nature in general is necessarily accompanied by the transcendental laws of nature: neither can exist without the other. Though the possibility of Nature in general is due to transcendental laws (B165; A216/B263), these laws have no application or significance apart from that nature (A160–161/B200–201). Both empirical and transcendental laws require some nature for their realization, just as both particular natures and Nature in general can exist only if accompanied by their associated laws.

I take the considerations presented in this section to show not just that proponents of the NA are on to something but that the NA should be extended beyond particular empirical laws to transcendental laws and noumenal laws (if there be any such). It is time to begin clarifying the NA and addressing objections to the extended version of it.

Just what are natures for Kant? How exactly should we understand the connection between laws and natures? I take the NA to posit a necessary connection between natures and their corresponding laws, such that the relevant laws always accompany the relevant natures and vice versa, but this is consistent with different models. Brian Ellis's dispositional essentialist account is one such model. For Ellis, there are various natural kinds, and there are properties that are essential to members of these kinds. The laws of nature derive their content and modal force from these essential

[27] See B165 and *MF* 4:472.

properties, some of which are dispositional. For example, there is a law to the effect that electrons have negative charge, and this arises from the fact that negative charge is essential to being an electron. Facts about laws, which have the force of metaphysical necessity, reduce to facts about essential properties of kinds of things (Mumford 2004: 106–109). This has led Stephen Mumford to object that the reductionist view is actually inconsistent with a governing account of laws. The problem is that if laws reduce to a fundamental base in the essential properties, then laws cannot be said to determine the base. As he puts it: "The laws merely ride on the back of these properties but, unlike jockeys on horses, cannot claim any credit for the direction the properties take" (Mumford 2004: 121). Is the Critical Kant working with a reductionist model like Ellis's that is open to this objection, or some other model? In Section 7.4, I attempt to answer this question after first trying to clarify the nature of a nature for Kant. In the conclusion, I then draw on this clarification to respond to two objections to the expanded NA.

7.4 The Nature of Natures and Laws

In the *Metaphysical Foundations*, Kant distinguishes between the term "nature" in its "formal" and "material" meanings. The latter refers to the sum total of all the objects of experience. Nature in its formal meaning, by contrast, refers to the "first inner principle of all that belongs to the existence of a thing" (*MF* 4:467). In this sense of the term, which is the one that matters for our discussion, we can speak of the natures of particular individuals and kinds of things (e.g., "fluid matter" and "fire" [A419n/B446n]), as well as of Nature in general.

Kant's definition, which is not unique to this text,[28] is rather obscure and itself in need of clarification. Since Kant tends to use the terms "nature" and "real essence" interchangeably, we can start by looking at what he says about real essences in the logic and metaphysics lectures.[29] Kant contrasts real essences with logical essences. The logical essence of a thing or kind of thing consists in the fundamental conceptual marks that are necessary for us to think of it. By contrast the real essence of a thing is "the first basic concept of everything that really and in fact belongs to the thing" (24:116). The properties that are essential for us to think of

[28] Cf. *Refl.* 18:180 [R5431] and 28:215–216.

[29] See 24:840 (from the Vienna Logic, transcripts of logic lectures that Kant gave around 1780) and 24:728 (from the Dohna-Wundlacken Logic, from the early 1790s).

something – those comprising its logical essence – need not be properties that are genuinely necessary for the thing. Real essences correspond to real definitions. Judgments about the properties belonging to, and resulting from, the real essence are synthetic rather than analytic (28:820).

But not everything that admits of a real definition and is the subject of synthetic judgments has a nature. Geometric figures satisfy these conditions, but Kant says they merely have essences rather than natures, because "in their concept nothing is thought that would express an existence" (*MF* 4:467n). What do we have to think in the concept of a thing for it to "express an existence" and thus for us to ascribe a nature to it? The answer, I think, is causal power. This I take to be the ability of a substance to ground the existence of determinations in itself – that is, positive states and behaviors – and to bring about changes in the determinations of other substances. The grounding in question is not logical, as is the case when a property follows from a concept in accordance with the law of noncontradiction, but rather "real." Remarks such as the following leave little doubt that Kant is thinking of fundamental causal powers as the "first inner principle[s] of the existence of a thing":

> It [i.e. the nature of a thing] concerns power and activity, the essential power is therefore the nature of the thing. E.g. The nature of quicksilver must contain the real ground of all of its consequences, i.e. the power, e.g. weight, fluidity, mobility. (28:49)

> The general real ground of the determinations inhering in a thing is nature; thus, that through which according to a general law, it is determined what belongs to the predicates of its existence. (*Refl* 18:180 [R5432])

Attractive force, for example, is one of the powers included in the nature of matter. This force is the ground for the fact that all matter exists with the following determination: it falls when it is dropped. As for quicksilver, Kant does not pretend to know the fundamental powers that belong to its nature, but whatever they are, they are the ground of the existence of its various states and of the changes it brings about in the states of other substances. As for Nature in general, there are no particular powers that belong to it, since as something common to all particular empirical natures it is nonspecific, but I would suggest that the concept of it includes the idea of causal power in general.

Let's turn to the issue of the connection between natures and laws. In the *Metaphysical Foundations*, Kant says that "the word 'nature' already carries with it the concept of laws, and the latter carries with it the concept of the necessity of all the determinations of a thing belonging to its existence" (*MF* 4:468). Putting this together with the previous points, I understand laws of nature to be rules that dictate the manner in which determinations must follow from the powers included in the nature of a thing. Laws do this by determining how the powers are exercised. For example, the law of universal gravitation dictates that bodies will attract one another to a degree directly proportional to their mass and inversely proportional to the square of their distance. The transcendental laws do not necessitate the existence of particular determinations, but they do provide some general constraints on the manner in which powers are related to determinations: for example, no change in a determination can occur without the exercise of some power of a substance.

So understood, laws are not themselves powers or forces – the law of universal gravitation is not what propels the book to the floor when it is dropped. Instead, the laws are, as it were, operating instructions included in the nature of a substance along with its powers. All particular natures have the transcendental laws that go along with Nature in general as very general operating instructions, along with a set of particular laws. Natures cannot exist without the laws that regulate the operation of the powers associated with them, and these laws cannot change without the natures being other than they are (whether it be a particular empirical nature or Nature in general). But laws likewise cannot exist without natures, since they presuppose powers whose efficacy is in need of regulation; they lack any inherent powers. The notion of a law of nature without a corresponding nature makes as little sense as a nature without a law.

In these respects, Kant's account of laws of nature is bottom-up. Is the account reductionist like Ellis's? I don't think so. For Ellis, the content and governing power of laws arises out of their essential properties, some of which are dispositional properties, similar to Kant's causal powers (Ellis 2001: 4–5). But Kant doesn't view the causal powers as giving rise to laws, since the latter have the job of directing the former; while the laws and causal powers are posited together with the nature of a given substance, and always go together, they are distinct aspects of it. Kant's position is thus not vulnerable to Mumford's objection.

If the content and governing power of laws is not reducible to features of the causal powers included in the empirical natures, where do they come from? I offer the following as a somewhat speculative answer to this difficult question. There is considerable evidence that

Kant thinks of all laws of nature as having a priori grounds. This was one of the key motivations for the DA. Now, the DA assumes that the a priori grounds in question must be epistemically accessible to us, since in the case of empirical laws, we must be able to use them to deduce the empirical laws from the empirical data. This led to the unfortunate result that empirical laws that cannot be deduced in this way – of which there is reason to think there are many – cannot be laws. But how can the NA itself make sense of Kant's insistence on the role of a priori grounds?

My suggestion is that we need a distinction between different types of a priori grounds of laws – ones that are epistemically accessible to us and ones that are not. What unites such grounds is that they are conditions of the possibility of the empirical natures associated with the laws. The complete set of a priori grounds for a given empirical nature will provide the full explanation for the nature and the corresponding laws being as they are. For such grounds to be epistemically accessible to us, they will typically have to be subjective conditions of our experience of the empirical nature in question, involving aspects of our pure understanding and/or sensibility. But laws can also have grounds of possibility that have nothing to do with us as subjects, in which case, while we might perhaps be able to see what the rules associated with them are, we can't understand why the content of the rules must be the way it is, and why things with that nature must behave in that fashion. But this does not foreclose the possibility of another being having a priori insight (working with the grounds of possibility) to grasp what we are missing.

Consider the transcendental laws, for example, the causal principle. The content of this rule and its necessity is comprehensible to us because it is, or can be deduced from, subjective conditions of the possibility of our experience of Nature in general. Something similar holds of the laws of mechanics. The content and necessity of these laws has its basis in subjective conditions of the possibility of our experience of the empirical nature of matter (the object of the empirical concept of matter). In particular, they are based on principles of the pure understanding (the transcendental laws of nature) and on aspects of our pure sensibility, particularly our ability to mathematize various aspects of the content of the empirical concept of matter. For this reason, we can see why it belongs to the nature of matter to act according to these particular mechanical laws and no others.

The case is different for other particular empirical laws, like the laws of chemistry. Kant is famously skeptical of our ability to come to know chemical laws because of their nonmathematical character (*MF* 4:471). I take it this is because their lack of mathematical content is an indication that the laws have their basis in a priori conditions of the possibility of the chemical natures in question to which we lack epistemic access. Such conditions would presumably involve features of the noumenal world (e.g., the noumenal natures that ground those chemical natures, and/or aspects of a divine understanding's manner of cognizing those natures). Though an a priori justification-cum-explanation of the chemical laws exceeds our abilities, it is not necessarily beyond the cognitive powers of another being (like God).[30]

7.5 Conclusion

In this chapter, I have tried to argue for, extend, and clarify the Necessitation Account. The NA was originally offered as an account of the particular laws of nature. Its core epistemic thesis is that particular laws are not necessarily knowable by us; its core metaphysical thesis is that Kant conceives of particular laws in terms of particular empirical natures, the positing of which necessitates certain regular behaviors. I argued that transcendental laws of nature, as well as noumenal laws (if there are any such) are similarly associated with natures. In the case of transcendental laws, the nature in question is Nature in general, which might be thought of as a completely general nature common to and contained in all particular empirical natures. The positing of this nature brings with it the transcendental laws and indeterminate causal power associated with it. In this respect, Kant has a uniform metaphysics of laws of nature.

The relationship between laws and natures was not initially clear. As I understand it, Kant holds that natures cannot exist apart from their associated laws, and those laws cannot exist apart from (must be realized within) their associated natures. This is the sense in which laws are inherent in natures and Kant's model of laws is bottom-up. However, the Critical Kant does not endorse a reductionist model like Ellis' dispositional essentialism. Laws do not reduce to the causal powers that define a given

[30] The law of universal gravitation is yet a third kind of case. Kant thinks that universal attraction is essential to matter and that this property is "comprehensible a priori" (*MF* 4:518). In addition, he clearly thinks we are justified in believing that masses attract one another in a manner inversely proportional to the square of the distances and directly proportional to their masses. But as for *why* the content of the law is precisely thus, i.e., why attractive force behaves in precisely this way, I take it the reason lies in a priori noumenal conditions of the possibility of matter to which we lack cognitive access.

nature but are a separate aspect of that nature, albeit one no more detachable from the nature than the causal power is. The law of a given nature, I suggested, acquires its content and necessity from the a priori grounds of the possibility of that nature. These grounds are not always epistemically accessible to us.

Let's close by addressing two objections to the expanded NA. The first concerns the apparent inconsistency between what it says about transcendental laws and Kant's Legislation Thesis. Kant's account of transcendental laws might seem to be a paradigmatic top-down model, with the human understanding playing something like the role that the divine will plays on the Occasionalist model of laws: superimposing laws onto a natural order that is not inherently lawful.[31] I think this is the wrong way to think about transcendental laws. Indeed, these laws have their source in the pure understanding. But it is also true that the transcendental laws are part of the identity conditions for all empirical natures; those natures would not be the natures that they are, would not be empirical natures, without conforming to the transcendental laws. The transcendental laws, in turn, exist as laws only insofar as there exists a Nature in general for them to inform. The Legislation Thesis is consistent with a bottom-up model of laws of this sort, whereby laws and their associated natures always exist together.

The second objection concerns the apparent inability of the NA to make sense of the manner in which particular laws of nature do not simply have an a priori ground but are grounded at least in part in transcendental laws. One passage indicative of this commitment was discussed in the context of the DA: "all laws of nature stand under higher principles of the understanding, as they only apply the latter to particular cases of appearances" (A159/B198). For Friedman, this is evidence of a relationship of epistemic dependence between the transcendental laws and the particular laws, such that particular laws can be deduced from the transcendental laws (and sometimes mechanical laws) in conjunction with the empirical data. What can the extended NA say about this? The reason that particular laws of nature stand under the transcendental laws is that the particular empirical natures associated with those laws all have in common a most general nature, Nature in general. In line with this, Kant speaks of the particular natures as "modifications" of the transcendental concept of nature, which corresponds to Nature in general (*CJ* 5:179). Insofar as this

[31] I take it this is the picture Ellis (2001: 1) has in mind when he groups Kant with proponents of a top-down model of laws.

Nature is governed by inherent transcendental laws, all of the particular empirical natures that are its specifications are governed by these laws as well. For this reason, all particular laws are subject to the general constraint that they do not violate the transcendental laws (B165). But this does not mean that we need be able to deduce all particular laws from the transcendental laws, even with all available empirical data, because in many cases these particular laws derive from conditions of possibility to which we lack epistemic access.

CHAPTER 8

Grounds, Modality, and Nomic Necessity in the Critical Kant

Michela Massimi

8.1 Nomic Necessity and the Kantian Problem of Inference

The natural world is characterized by a lawful order and harmony. Tides occur at regular intervals, the lunar cycle is 29.5 days, and chemical substances have atomic numbers that explain a variety of their physico-chemical properties. Laws of nature seem to govern the natural world and to explain its robust regularities. Given Newton's law of gravity, and given the masses of the Earth and the Moon and their respective alignment, high tide *necessarily* follows. Given the laws of chemistry, the properties of aqua regia, and of gold, if thrown in aqua regia, gold *necessarily* dissolves.

Nomic necessity suggests that laws of nature go beyond Humean regularities and capture a robust modal thought: if C had been the case, *necessarily* E would have been the case. Modality has traditionally been invoked to distinguish between laws of nature and accidentally true universal generalizations such as "All fruits in Smith's garden are apples," where it is not a law of nature that were a seed thrown in Smith's garden, necessarily an apple tree would grow out of it.

How did Kant explain lawlike claims such as "Were the Earth and Moon aligned in a given way, high tide would *necessarily* occur"? In the *Critique of Pure Reason* Kant famously stated that the categories are concepts that prescribe laws a priori to nature "as the sum total of all appearances (*natura materialiter spectata*)" (B163). He then presented "a riddle" (I am henceforth going to call it the *Kantian problem of inference*). Namely, how can nature possibly follow the a priori laws that the understanding prescribes to it?

Since the categories are not themselves derived from nature, nor do they follow nature's patterns (otherwise they would be empirical and not a priori), it would seem preposterous to expect them to have any purchase on nature itself. This is the "riddle," as Kant outlined it. Nonetheless, we are told that the faculty of understanding *prescribes laws to nature.* Let us state the problem more precisely. The Kantian problem of inference takes the following form:

(I) All alterations occur in accordance with the law of the connection of cause and effect (Second Analogy).
(II) Event of type A (e.g., alignment of the Moon and the Earth) causes event of type B (e.g., high tide occurring – empirical causal law).[1]
(III) Event A_1 causes event B_1 (e.g., this particular Moon–Earth alignment today causes high tides at Cramond Island at 11:35 a.m. – instantiated empirical causal law).
(IV) A_1 occurs.
(V) B_1 *necessarily* follows (necessity of effects, via I, II, and III).

For Kant, I argue, the causal connection at work in lawlike claims seems to follow this inferential pattern, which takes us from pure principles of the understanding such as (I) to specific events (i.e., V) via type–token empirical causal laws (II and III). The Kantian problem of inference[2] is the problem of how to proceed from a general premise such as Kant's Second Analogy of Experience (I) to the conclusion that *necessarily* high tide occurs at Cramond Island at 11:35 this morning (V), given that the Earth and Moon are aligned in a particular way today (IV).

The Kantian problem of inference is twofold. First, if the *understanding prescribes laws to nature*, lawfulness seems to be restricted to phenomena as objects of possible experience, and it does not extend to things in themselves. Thus, it would seem that the understanding prescribes laws to nature in a purely formal sense (qua *natura formaliter spectata*), that is, when nature is considered with respect to our formal conditions of the possibility

[1] There is a debate on how to interpret the Second Analogy (i.e., with a weak reading, "every event has a cause," or a strong reading, "same event, same cause"). Here and in what follows, I understand the Second Analogy as implying the stronger reading "same event, same cause" (primarily defended by Guyer 1987).

[2] The terminology "problem of inference" is coined by van Fraassen 1989 in his critical treatment of David Armstrong's Necessitarian account of laws. In van Fraassen's use, the problem of inference is the problem of how to move from a necessitation relation between universal properties such as *N* (*F-ness, G-ness*) to the necessitation relation that is supposed to hold at the level of particulars *N* (*Fa, Ga*); *N* (*Fb, Gb*); and so on. I borrow the terminology, although – it should be clear from the passage above – the Kantian version of the problem of inference concerns how to go from a kind of necessity captured by the Second Analogy in its determination of the connection of appearances to the necessity that is supposed to hold in nature between events that we come to know via the Analogies of Experience, and more in general via the categories of the understanding. Thus, the Kantian problem of inference is even more pressing than its Armstrongian counterpart because it is the problem of explaining how to infer the necessity of particular lawful events in nature from the necessary determination of the connection of appearances that our faculty of understanding makes possible.

of experience. But the understanding would not prescribe laws to nature in any material sense, that is, in the sense of delivering on the promise of answering the question about how laws of nature can have a purchase on *nature itself.* Let us call this the *metaphysical quandary* of the Kantian problem of inference. The metaphysical quandary is not going to dissolve by appealing to transcendental idealism or the objective validity of the categories of the understanding. The thought of a purely formal nature being dependent on us, so that the necessity of effects would almost analytically follow from the causality expressed by the Second Analogy, is of course tempting in this context. But if this were indeed Kant's final and considered answer to the problem of inference, it would be bad news.

Recall our key question is: how did Kant explain lawlike claims such as "Were the Earth and Moon aligned in a given way, high tide would *necessarily* occur"? If it turns out that Kant's considered answer to this question (and similar ones) is that the understanding prescribes laws to nature qua sum of appearances that are (precisely because appearances) dependent on us, then it might well be possible that no tide will actually occur at the next Earth–Moon alignment (or that no tide would in fact ever have occurred at any Earth–Moon alignment) had it not been for our faculty of understanding imposing the Second Analogy on a relevant group of appearances.

Yet our key question has metaphysical import. It enjoins us to think that were the Earth and Moon aligned in a given way, high tide would *necessarily* occur (and indeed, it would necessarily occur even if there were no human beings advancing knowledge claims about tides and their lawlike occurrences). This is what I call the *metaphysical quandary* of the Kantian problem of inference.

Second, in the relevant passage of the *Critique of Pure Reason*, Kant is adamant that "Particular laws, because they concern empirically determined appearances, **cannot** be **completely derived** from the categories, although they all stand under them. Experience must be added in order to come to know particular laws **at all**; but about experience in general, and about what can be cognized as an object of experience, only those *a priori* laws offer instruction" (B165). Thus, it is not the case that the empirical laws of magnetism, or this particular causal law about tides (II), or any other, can be derived from the Second Analogy of Experience.

Instead, as Kant puts it, empirical causal laws simply "stand under" the categories and their principles, for example, the Second Analogy. I suggest that we see (II) as an instantiation of the general cause–effect template captured by the Second Analogy (I), pretty much as (III) is an instantiation of (II), in turn. Experience plays a role in our knowledge of these different instantiations in a way that we need to clarify in the rest of this chapter. How can we then *come to cognize* the *necessity* of effects (V) as following from the general principle of causality expressed by (I)? Necessity attaches to the principle of causality via the Postulates of Empirical Thinking in General, which do not add anything to the content of the categories but only express "*the cognitive power whence it* [concept] *arises and has its seat*" (A234/B287). It is then unclear how one can *claim to cognize the necessity* of empirical causal laws, which can be derived neither from the understanding a priori nor from experience a posteriori. Let us call this the *epistemological quandary* of the Kantian problem of inference.

To answer both quandaries, I suggest we turn our attention to Kant's discussion of grounds and modality and delve into Kant's lectures on metaphysics so as to acquire a better grasp of his considered view on the nomic necessity of the laws.[3] My final goal is to offer an account of the necessity of the laws of nature that I hope can make some progress on the Kantian problem of inference.

8.2 Three Kinds of Grounds and Three Kinds of Necessity in Kant

In this section, I turn to Kant's lectures on metaphysics to clarify why, in my view, Kant's considered answer to the problem of inference should be searched for in his multifaceted notion of ground and consequence. I clarify three different kinds of nomic necessity that Kant saw at play in different kinds of laws, each respectively relying on a different notion of ground, qua conceptual ground ("*ratio cognoscendi*"), qua ground of being ("*ratio essendi*"), or qua ground of becoming ("*ratio fiendi*"). Only the latter notion captures cause–effect relations at play in empirical causal laws, I argue.

In *Metaphysik Mrongovious* (written around 1782–1783),[4] Kant called the relation of ground and consequence a connection "*nexus*" and distinguished between analytic or logical connection – as that "according to

[3] In a seminal paper on Kant's answer to Hume, Paul Guyer (2003a) has argued that since for Kant every judgment has a relation (it is either categorical or hypothetical), in the case of hypothetical judgments of the form "if *p* then *q*" Kant deployed the logical relation of *ground* and *consequence* without either clearly distinguishing it from the relation of cause and effect or including it in the table of categories. In what follows, I expand on Guyer's distinction between grounds and causes so as to achieve a better grasp of both the metaphysical and epistemological quandary at stake in the Kantian problem of inference.

[4] Kant did not leave his own lecture notes on metaphysics; all we have left is notes taken by his students. What is striking about these notes is the overlap of themes over the years and in notes taken by different students, so we can be confident enough that they reflect Kant's genuine and sustained thoughts on the matter.

the principle of identity" – and what he called synthetic or real connection – "if it is not according to this principle" (29:807). Connections rest on grounds, which in turn can be either logical grounds or real grounds,[5] depending on whether something is posited according to the principle of identity or not.[6] For example,

> extension is a ground of divisibility, the latter is posited through the former according to rules of identity, – but: every body has attractive power, here the latter is posited through the former (body), but not according to the rules of identity, and this connection <*nexus*> is real, the former logical. It is possible to cognize a real connection <*nexus*> only *a posteriori*. (29:807)

Mark these words. Kant starts with a general logical definition of ground as "that which, having been posited, another thing is posited determinately, the consequence is that which is not posited unless something else is posited" (29:808). Logical grounds underpin logical connections, that is, connections "according to the principle of identity." I suggest reading logical grounds as cognitive grounds, or *rationes cognoscendi*, as when we infer divisibility from extension (or freedom from the moral law, see *CPrR* 5:004).[7] More precisely, I suggest that we understand logical grounds as *conceptually* determining grounds (qua *ratio cognoscendi*) that deliver rules for inferring a conceptual consequence once a conceptual ground is posited. Using contemporary language, one might say that, for Kant, to be *conceptually necessary* is to be true in virtue of the nature of the concepts involved. Thus, the concept of a body's divisibility is grounded in the concept of a body's extension. Logical grounds qua *conceptually* determining grounds are inferentially deployed in lawlike judgments of the form:

(i) Necessarily, if bodies are extended, they are divisible.

[5] For a seminal analysis on real grounds, see Watkins 2005 and Warren 2001.

[6] Kant develops this distinction further in *Metaphysik* L_2 (dating back to 1790–1791), where he cashes out the distinction between logical grounds and real grounds as the distinction between "the relation of cognition, how one is inferred from the other" and the analogous relation in metaphysics where "ground belongs under the concept of causality" (28:548). In both cases, grounds bring along with them necessary connections with their consequences: "if I posit the ground, then a consequence must follow necessarily. A ground is that whereby, when it is posited, another thing is determinately posited <*ratio est id, quo posito determinate ponitur aliud*>" (28:549).

[7] "Now we have a criterion of the ground, namely: that which, having been posited determinately, another is posited <*quo posito determinate, ponitur aliud*>. Determinately <*determinate*> means according to a general rule <*secundum regulam generalem*>. Every ground gives a rule, therefore the connection <*nexus*> of the ground and the consequence is necessary. (Logical ground is a cognition from which another follows according to a rule)" (29:808).

The necessity captured by (i) is what Kant calls the necessity according to the principle of identity for logical connections. We may want to call it *conceptual necessity*. That Kant saw (i) as expressing a conceptually necessary truth is clear from his pre-Critical defense of John Keill's theorem for the infinite divisibility of the space that bodies fill (with related geometrical demonstration) in *Physical Monadology* (*PM* 1:478).[8]

Conceptual necessity should not be confused with the necessity at play in theoretical identity statements, which take syntactically a form akin to (i) but express a very different kind of necessity:

(ii*) Necessarily, if something is water, then it is H_2O.
(ii**) Necessarily, if something is gold, then it has atomic number 79.

Or closer to Kant:

(ii ***) Necessarily, if a body has repulsive force, it is impenetrable.

Theoretical identity statements are *necessary a posteriori,* according to a very influential contemporary Kripkean account, because they deliver knowledge about the essence of things, and such knowledge is a posteriori (i.e., it is the result of a scientific discovery). Thus, in this sense the necessity of theoretical identity statements is not the necessity with which Kant's logical grounds qua *rationes cognoscendi* inferentially lead to their consequents: an experimental discovery is required to come to know that water is H_2O or that matter is impenetrable in virtue of its repulsive force.[9] The necessity of theoretical identity statements is the necessity that attaches to the essence of water, the essence of gold, or the essence of matter, and for which, I suggest, we need to resort to another tool in Kant's arsenal, namely, grounds of being, or *rationes essendi.*

Kant's discussion of grounds of being is scant and not very illuminating. *Ratio essendi* sits uncomfortably between logical grounds and (proper) real grounds as grounds of becoming (*ratio fiendi*). In what follows, I give my own (arguably fallible) reading of Kant on *grounds of being* and *grounds of*

[8] The notion of *ratio cognoscendi* features already in the 1755 pre-Critical text *New Elucidation* but in a slightly different way. For it is there identified with what Kant calls "consequently determining reason," which is one of the two possible expressions of the Leibnizian principle of sufficient reason (or better, principle of determining reason, as Kant calls it), rather than just a logical ground as in the aforementioned quotes from the much later, Critical-period metaphysics lectures.

[9] Odd as it might sound to our contemporary ear, Kant regarded impenetrability as an essential property of bodies as much as we would now regard atomic number 79 an essential property of gold. For Kant matter was defined as that which fills a space and resists penetration. Hence the sentence "Bodies are impenetrable" would enjoy – I contend – a status that is akin to our contemporary theoretical identity statements about gold and water.

becoming, and how I see both of them at work in two further kinds of nomic necessity that Kant seems to be subscribing to.

If one is asking not whether a concept is the ground for inferring another concept, but rather "how a thing is the ground of other things," that is, how it is possible that "if one thing is posited, the other can be determined by it" (29:809), one must turn to metaphysics rather than logic. Thus, real grounds (properly speaking) capture an entirely different class of lawlike claims: those where, if a ground is posited as a *cause*, a consequence follows as an *effect*, and not just logically or conceptually but *existentially*.[10] Kant calls a real ground qua cause "the ground of becoming <*ratio fiendi*>," not to be confused with the ground of being *ratio essendi* as "the ground of that which belong to a thing considered according to its possibility, for example, the three sides in the triangle are the ground of the three corners" (ibid.).

I take Kant's *ratio essendi* to denote essential properties of things. For example, it is an essential property of triangles to have three sides, and once that is posited the possibility of three corners is also posited because they are both properties that pertain to the essence of triangles.

Using modern notation,[11] one might be tempted to write Kant's *ratio essendi* as follows: $\Box_x p$, namely, *p obtains in virtue of what it is to be x*. Grounds of being as *ratio essendi* capture not just geometrical claims such as the one above about triangles. But – I contend – they capture any lawlike claim that pertains to the *possible determinations* of the essence of a thing. In *Metaphysik* L_1 (one of Kant's early sets of lecture notes from the mid-1770s) Kant distinguished between the *nature* of a thing and the *essence* of a thing:

> The inner ground [of that] which belongs to the actuality of a thing is *nature*, but what belongs to the possibility and to the concept of the thing is *essence*. A triangle has no nature, for it is no actuality, but rather only shape, thus in all of geometry there is no nature ... The essence of a body is that which belongs to its concept; but nature [is that] by which all phenomena can be explained. What is general in the nature of bodies, what contains the

10 "The two concepts of ground and consequence are logical but not transcendental. Cause and effects are things. Cause is that out of which the existence of another follows. Existence is not at all discussed in logic" (29:809).

11 For this notation, see Fine 1994, 2012; Correia 2011; and Rosen 2010. In modern wording, one would read the grounding operator $\Box_x p$ as follows: "it lies in the nature of *x* that *p*." But since Kant – as I am going to clarify below – sharply distinguished between the *essence* of things and the *nature* of things, to avoid terminological confusion between Kant's vocabulary and the modern one, let us read the grounding operator $\Box_x p$ as "*p obtains in virtue of what it is to be x*." See Chapter 7 in this volume for additional textual evidence about Kant's view on the "nature" of a thing.

> principle of all phenomena, is very little, namely impenetrability, connection, and shape. (28:211)

In *Metaphysik* L_2 (written in 1790–1791, around twenty years after *Metaphysik* L_1) Kant further elucidates the difference between a *logical essence* as "the first ground of all logical *predicates* of a thing" and a *real essence* as "the first ground of all *determinations* of an essence" (28:553). While logical essences require the analysis of all predicates that lie in any given concept, real essences are found through the principles of synthesis and are "the first inner ground of all that which belongs to the matter itself":

> The real essence is not the essence of the concept, but rather of the matter. E.g., the predicate of impenetrability belongs to the existence of a body. Now I observe through experience much that belongs to its existence; e.g. extension in space, resistance against other bodies, etc. Now the inner ground of all this is the nature of the thing. We can infer the inner principle only from the properties known to us; *therefore the real essence of things is inscrutable to us*, although we cognize many essential aspects. We become acquainted with the powers of things bit by bit in experience. (28:553).

Thus, although we can never know the real essences of things, we become familiar with "the powers of things bit by bit" in experience. We come to cognize by experience essential properties or *determinations that belong to the existence of a body*, for example, that bodies resist penetration from other bodies and attract other bodies too. From these properties or determinations known to us, we infer the "inner ground," or what Kant calls the "nature of the thing." Thus, from fermentations and chemical and optical phenomena that Kant was well familiar with and to which he repeatedly referred in *Universal Natural History*, *On Fire*, *Physical Monadology*, and again *Metaphysical Foundations*, we come to infer that there must be a repulsive force as a power that allows bodies to resist penetration, saltpeter to burn in the bowels of the Sun, winds to form when there is a decrease in the expansive force of the air, and so on.[12]

In a quasi-Kripkean fashion, a theoretical identity statement such as (ii***) would then display a kind of necessity, which is better called *metaphysical necessity* because it is grounded in the essential determinations or powers of matter. I take Kant's "ground of being" qua *ratio essendi* to support two (weak and strong) modal claims:

[12] For details about Kant's take on repulsive force in these phenomena, see Massimi 2011. For an analysis of Kant's notion of "nature" of things and its relation to explanatory laws, see Ameriks's (2012) treatment in the chapter "Kant and the End of Theodicy."

> ***Ratio essendi* weak modal claim:**
> Given an essential property *F* that obtains in virtue of what it is to be *x* (qua ground of being), positing *x* metaphysically grounds the possibility of *F*.

For example, given that having three corners is an essential property that obtains in virtue of what it is to be a triangle *x* (qua ground of being), whenever we posit *x*, we posit also the possibility of having three corners. Or, to use a different example, given that resisting penetration is for Kant an essential property that obtains in virtue of what it is to be matter *x* (qua ground of being), by positing *x* the possibility of impenetrability *F* is also posited.[13]

Kant would presumably endorse an even stronger claim whereby, whenever the ground of being for triangles is posited, *necessarily* the possibility of the essential property of three corners is also posited. Similarly, whenever the ground of being for matter is posited, *necessarily* the possibility of the essential property of impenetrability is posited with it:

> ***Ratio essendi* strong modal claim:**
> Given essential property *F* that obtains in virtue of what it is to be *x* (qua ground of being), it is necessary that positing *x* metaphysically grounds the possibility of *F*.

I take the *Ratio essendi* strong modal claim to be at work in theoretical identity statements such as "Water is H_2O," or "Gold is element with atomic number 79," or, closer to Kant, "Triangles have three corners" or "Bodies are impenetrable." Theoretical identity statements are *necessary a posteriori* because the fact that it is in the essence of material bodies to resist penetration *metaphysically grounds* the possibility of impenetrability, and necessarily so (according to the *Ratio essendi* strong modal claim).

Things get complicated with real grounds qua *causes* or *rationes fiendi*. In light of Kant's aforementioned distinction between essence and nature, Kant seems to be suggesting that if something obtains in virtue of what it is to be a real ground qua *ratio fiendi*, then it is not just the possibility of some essential property *F* that is grounded in the obtaining of the ground *x* (and necessarily so). Instead, by positing the real ground, a whole host of lawful

[13] The *ratio essendi* weak modal claim does not contend that the possibility of *F* is included in actuality of *x* (or better, in the actuality of what it is to be *x*). It simply tells us that *if F* obtains in virtue of what it is to be *x*, then positing this essentialist claim metaphysically grounds the possibility of *F*. Whether or not there exists a *ground* in actuality that indeed metaphysically grounds the possibility of *F* falls outside the scope of what the *ratio essendi* weak (and strong) modal claims can ever teach us. As we will see shortly, telling us whether or not there exists such a *ground* is the job of the *ratio fiendi*'s first modal move.

phenomena are *causally grounded*, and necessarily so. This necessity of synthetic connections, Kant adds, is baffling. No human reason can comprehend it, and it cannot be grasped by experience either: "That which the real ground contains of something is called *cause*. I can *not* comprehend the concept of real ground from experience; for it contains a necessity" (28:549). What kind of necessity do real grounds (*rationes fiendi*) deliver?

It is not the conceptual necessity licensed by logical grounds, because the principle of causality bridges antecedents and consequents in lawlike claims *synthetically*, that is, without the concept of the consequent being already included in the concept of the ground. It is not metaphysical necessity either, licensed by grounds of being or *rationes essendi*, because the necessity of "grounds of becoming" undergirds lawlike claims that are *not* confined to the possible essential determinations of things (and necessarily so; e.g., matter being impenetrable, and necessarily so). Instead, the necessity of "grounds of becoming" undergirds lawlike claims that go from the *actual* obtaining of something qua *cause* to the *actual* obtaining of something else qua *effect in nature*.

Real grounds qua *rationes fiendi* then capture modal connections, whereby the real ground is a *causal* ground for the necessary occurrence of the consequent: facts featured in the antecedent of the "if . . . then" claim are *causal grounds for* the consequent facts obtaining. This kind of necessity is neither conceptual nor metaphysical. It is *natural*, I contend.[14] For real connections to be *naturally necessary* is for them to be true in virtue of what Kant calls the "nature of the things" (and not the essence of the things, since for Kant *essences* give only the grounds for the possibility of things, but *natures* give grounds for the actuality of things). For example:

(iii*) Necessarily, if the Earth has attractive force, then the Earth has a spherical shape.

(iii**) Necessarily, if air has repulsive force, then air produces winds.

These statements are necessary because – to Kant's eyes – the consequent obtains in virtue of positing attractive or repulsive forces that *cause* the Earth to acquire its spherical shape, or *cause* air to produce winds, respectively. And attractive and repulsive forces are essential properties or powers that belong to the *nature* of matter. While the necessity associated with logical grounds is what I have called *conceptual necessity*, and the necessity

[14] Natural necessity is not one and the same as metaphysical necessity, for the latter subdivides into other nonnatural kinds of necessity, such as normative necessity (for a classic contemporary treatment, see Fine 2002).

associated with grounds of being, or *rationes essendi*, is what I have called *metaphysical necessity*, I contend that the necessity that is to be found in most empirical causal laws is the *natural necessity* typical of real grounds qua *rationes fiendi*.

Equipped with this threefold distinction of grounds qua *ratio cognoscendi*, *ratio essendi*, and *ratio fiendi*, and their respective kinds of necessity, we can go back to the *metaphysical quandary* associated with the Kantian problem of inference and provide an answer to it on Kant's behalf.

8.3 Solving the Kantian Problem of Inference: *Causal Grounds* and Their Triple Modal Move

Recall the metaphysical quandary. Kant's famous claim about the understanding prescribing laws to nature seems prima facie to confine lawfulness to phenomena as objects of possible experience. Thus, it would seem that the understanding prescribes laws to nature in a purely formal sense (qua *natura formaliter spectata*), but not in any material sense, that is, in the sense of delivering on the promise of nomic necessity that would easily answer the question about how laws of nature can possibly have a purchase on *nature itself*.

The analysis in the previous section makes us well equipped to go back to this quandary and offer a possible solution to it. Laws of nature – I argue on Kant's behalf – have a purchase on *nature itself* because they express the way in which once real grounds are posited, consequences necessarily follow. Real essences are inscrutable to us (hence the lawfulness of things in themselves goes beyond the boundaries of our experience). Yet we do cognize many essential properties or determinations of things, or what Kant also calls "the powers of things," such as attractive and repulsive forces. These powers of things, or essential determinations that belong to the existence of matter, are rooted in real grounds qua *causal grounds* for the consequent facts obtaining.

Under the broadly dispositional essentialist reading that I am going to suggest on Kant's behalf,[15] Kant would be concerned with how some essential properties or powers – grounded in the "nature of things" qua real grounds (*rationes fiendi*)[16] – result in laws of nature that have a

[15] This reading follows up on Massimi 2014. For a further dispositional essentialist reading of Kant and laws, along lines that are slightly different from mine, see Chapter 7 in this volume.

[16] On my reading here, the "natures" of things (or *rationes fiendi*) are not one and the same as the causal powers of things. Instead they ground the latter, in the sense of providing *metaphysical grounding* for

purchase on nature itself via a triple modal move. Kant calls these essential properties "*essentialia*" (28:553) or, alternatively, "the powers of things," and he takes attractive and repulsive forces as the two main examples. He is committed to the view that these essential properties or powers are not quiddities, whose identity is primitive and cannot be further explained or grounded in anything else.[17] Instead, he seems to suggest that those essential properties or powers have an "inner ground" or what he interchangeably calls a "real ground," which is in turn the "nature of the thing" (I am going to call it *z* to distinguish it from ground of being *x*), whose role – I take it – is to *metaphysically* ground these essential properties.[18] Hence, Kant's view is essentialist in bringing in the "nature of the thing" with the following first modal move, which I am going to call the *Ratio fiendi* first modal move:

> ***Ratio fiendi* first modal move:**
> There exists a real ground *z* such that essential property or power *F* obtains in virtue of what it is to be *z*. Hence, by positing *z* as a real ground, essential property, or power, *F* is also determinately posited.

For example, there exists a real ground *z* of, say, matter such that all material bodies have the power of resisting penetration.[19] This first modal move captures the metaphysical necessity of essential properties or powers *F*, which for Kant are not quiddities but have an inner ground/real ground

essential determinations or powers of things. Hence the first *essentialist* component of my triple modal move here.

[17] Thus, Kant's view differs from what is now called categoricalism, i.e., the view that underlies contemporary Humean metaphysics à la Lewis no less than Necessitarian accounts such as Armstrong's. Categoricalism maintains that the fundamental natural properties are categorical, rather than dispositional: their identity is defined by a quiddity, not by their possible causal roles (as in the case of dispositions). Categorical properties may be instantiated in nature with very different causal roles, without yet losing their identity, which is secured by quidditism. Hence the laws of nature where categorical properties feature are metaphysically contingent. That is not the case with dispositional essentialism: a dispositional essential property is always instantiated with the same causal role (no matter what the actual manifestations are) because it is identified by such a role (and not by some nonexplicable quiddity). For an excellent introduction see Choi and Fara (2016).

[18] I take it that both *rationes fiendi* and *rationes essendi* have the primary task of metaphysically grounding essential properties, but with two main differences. First, *rationes essendi* metaphysically ground the possibility of essential properties, whereas *rationes fiendi* metaphysically ground the actuality of essential properties (recall the difference Kant draws between essence and nature). Second, and related to my first point, the class of essential properties whose possibility is metaphysically grounded by *rationes essendi* is larger than the class of essential properties whose actuality is metaphysically grounded by *rationes fiendi* (e.g., the former includes geometrical properties such as triangularity, among others).

[19] I take Kant to be subscribing to this first modal move because he takes real grounds as grounds of actuality and not just grounds for the possibility of some essential properties (recall quotes 28:211 and 28:553).

z in what he calls the "nature of the thing." To reiterate: Kant does not identify real grounds (*z*) and essential properties (or powers) (*F*). Nor does he consider essential properties as sheer predicates that can be attributed to things. Real grounds qua *rationes fiendi metaphysically ground* the actuality (and not just the possibility) of essential properties or powers – this is what the *Ratio fiendi* first modal move tells us. But we are not simply interested in the metaphysical necessity with which essential properties or powers are actually posited, whenever real grounds are posited. What is distinctive and unique about real grounds qua causes are the further modal claims they make possible.

Indeed, not only is Kant endorsing the *Ratio fiendi* first modal move concerning essentialism about properties or powers. He is also endorsing – I contend – a further modal claim that explicates the properly *dispositional nature* of his essentialism.[20] Following dispositional essentialism as the view that an essentially dispositional property manifests a given behavior in the presence of the right stimulus condition,[21] I suggest that we understand Kant's argument as implicitly buying into the following dispositionalist modal move:

> ***Ratio fiendi* second modal move:**
> There exists a real ground *z* such that it is in virtue of *z* that each object *a* that has it also has the essential property or power *F* (as per the *Ratio fiendi* first modal move); and this fact, in turn, causally grounds the dispositional behavior *D* of any object *a*.

For example, the elasticity of material bodies $a_1 \ldots a_n$ is a disposition *D* that is causally grounded in the repulsive force as an essential power *F* that lies in the real ground *z* of any material body *a*.

Kant clearly saw the relation between real grounds, essential properties or powers, and their manifestations along these dispositionalist essentialist lines. Real grounds pertaining to the nature of things undergird powers that in turn *causally ground* dispositional behavior *D*. For example, repulsive force as an essential determination of matter is manifested in "the self-same elasticity and pressure of the air," which is in turn "of necessity the

[20] Here I fully endorse Barbara Vetter's 2012 distinction between (1) some fundamental properties are essential, and (2) some of these essential properties are dispositional. E.g., the essential property of having three sides for triangles is not dispositional, but the essential property of repulsive force for material bodies is dispositional.

[21] According to a sophisticated conditional analysis of dispositions, object *a* has a disposition *D* to manifest *M* in the presence of the right stimulus condition *S* iff *a* has some essential property (or power *F*) that would *cause a* to manifest *M* in the presence of *S* (for this conditional analysis of dispositions, see Lewis 1997). I return to the importance of *M* and *S* below in the *ratio fiendi* third move.

ground of the possibility of pumps, of the generation of clouds, of the maintenance of fire, of the winds, and so on. It is necessary that, as soon as the *ground* of even merely one of them be present, the *ground* of the others should also be present" (*OPA* 2:106, emphasis added). In other words, as soon as we posit repulsive force (*F*) as a power that lies in the nature of a thing (or in its real ground *z*), we will see *F* causing the dispositional behaviour of elasticity (*D*) that is at work in many effects wherever *z* is present (i.e., the generation of the clouds, the maintenance of fire, the production of the winds, and so forth).

Following a long tradition that goes back to Newton and Hales, among others,[22] Kant thought, for example, that the repulsive force was an essential power that belonged to the existence of air. Along similar lines in the *Metaphysical Foundations*, he declared, "The expansive force of a matter is also called *elasticity*. Now, since it is the basis [*Grund*] on which the filling of space rests, as an essential property of all matter, this elasticity must therefore be called *original*, because it can be derived from no other property of matter. All matter is therefore originally elastic" (*MF* 4:500). Similarly for attraction: "The *attraction essential to all matter* is an immediate action of matter on other matter through empty space . . . This original attractive force contains the very ground of the possibility of matter, as that thing which fills a space to a determinate degree, and so contains even [the ground] of the possibility of a physical contact thereof" (*MF* 4:512).

Finally, to complete Kant's response to the metaphysical quandary, we need a further modal claim that links dispositional essential properties (e.g., the elasticity due to the repulsive force or the action at a distance due to the attractive force) with their very many manifestations in nature. To achieve this goal, I suggest that we unpack the *Ratio fiendi* second modal move to allow for what Alexander Bird (2007), following Gilbert Ryle, calls "multi-track dispositions," that is, dispositions $D_{(S,M)}a$ that have various manifestation conditions *M* under slightly varied stimuli conditions *S*:[23]

> ***Ratio fiendi* third modal move:**
> There exists a real ground *z* for essential property or power *F*, which causally grounds the dispositional behavior *D* of any object *a* having *z* as

[22] For details, see Massimi 2011.

[23] Multi-track dispositions have more than one pair of manifestation *M* and stimulus *S* condition (although they must share the same notion of stimulus condition). Being elastic is an example of a multi-track disposition because it can have more than one pair. Its stimulus condition can be being released, being stretched, being pulled, being twisted. Its manifestation condition can be being inflated, being bounced back, being squidgy, being resistant, and so on.

its real ground and F as an essential power, so that in different stimuli conditions S, different manifestations M of disposition D occur.

Under slightly varied stimuli conditions $Sa_1 \dots Sa_n$, the repulsive force F – as a multiply instantiable essential power grounded in z – displays a multi-track disposition $D_{(S,M)}a$, that is, elasticity (depending on whether it is the elasticity released by the atmospheric air, the elasticity inherent in colliding balls, or the elasticity of a stretched spring, and so on). This multi-track disposition *necessarily* yields a variety of manifestations M. For example, Ma_1, that is, winds are produced when elasticity decreases and causes alteration in the equilibrium of the atmosphere;[24] Ma_2, that is, fire burns when elastic air is released by saltpeter in the bowels of the Sun;[25] Ma_3, that is, a spring recoils when its elastic material is being pulled; and so forth. What all these different phenomena share is a common real ground z in virtue of which an essential power F obtains (e.g., repulsive force), such that F causally grounds a multi-track disposition $D_{(S,M)}a$ (e.g., elasticity), which under different stimuli conditions necessarily brings about different manifestations M.

To sum up, the three modal moves I have suggested are designed to deliver a three-tier causal basis for multi-track dispositions in Kant's metaphysics. Real grounds are the first tier. Essential properties or powers F are the second tier. Multi-track dispositions are the third tier.[26] More to the point, the essential properties or powers of the second tier are not sparse categorical properties, because (1) Kant rejects quidditism and grounds these properties in real grounds (as per *Ratio fiendi* first modal move). Moreover, (2) Kant identifies these essential properties or powers with their causal roles in bringing about multi-track dispositions (as per *Ratio fiendi* second and third modal moves). As for real grounds themselves, they are neither categorical nor dispositional properties – indeed they are not properties at all but instead grounds for powers (and ultimately for the dispositions that supervene on them).

[24] In his 1756 *Theory of the Winds* the cause of the winds was itself identified with the "decrease of the expansive force by cold and vapors that reduce the elasticity of the air" (1:491).

[25] In *Universal Natural History*, Kant described elasticity or, better, "elastic air" as "capable of maintaining the most violent degrees of fire" in the atmosphere of the sun (1:325–326).

[26] According to an influential view by Prior, Pargetter, and Jackson (1982), dispositions have a causal basis, which is typically constituted by a set of properties. Whenever an object has this set of essential properties and is in the right stimulus condition, it is causally necessary that the object manifests dispositional behavior D. Critics have objected that if the causal work of necessitating the manifested behavior is done by the causal basis of properties, dispositions are causally inert and redundant. I hint at an answer to this objection below (at least in the context of my analysis of Kant's view).

If this analysis is correct, a new picture emerges about Kant's view on the necessity of laws. For multi-track dispositions are ultimately grounded on real grounds all the way down. Moreover, it is this three-tier structure of real grounds, essential properties or powers, and dispositions that ultimately cause (and necessarily so) the stimulus–manifestation pairs. The essential power of repulsion (in and of itself) neither causally explains nor necessitates that saltpeter should burn inside the Sun or that winds should form by decreasing the elasticity of the air. For any stimulus–manifestation pair, the three-tier system of real grounds, essential powers, and multi-track dispositions must be present.[27] And it is this three-tier system that ultimately makes laws of nature *necessary* and *part of a unified system of laws*, I contend.

We are now equipped to go back and solve the metaphysical quandary and explain why, on Kant's account, laws have a purchase on nature and can *necessitate* states of affairs, despite the fact that the understanding prescribes laws to nature in a purely formal way (qua *natura formaliter spectata*). How does the inference from (I) to (V) proceed?

(I) All alterations occur in accordance with the law of the connection of cause and effect (Second Analogy).
(II) Event of type A (e.g., alignment of the Moon and the Earth) causes event of type B (e.g., high tide occurring – empirical causal law).
(III) Event A_1 causes event B_1 (e.g., this particular Moon–Earth alignment today causes high tides at Cramond Island at 11:35 a.m. – instantiated empirical causal law).
(IV) A_1 occurs.
(V) B_1 *necessarily* follows (necessity of effects, via I, II, and III).

On the reading I have been suggesting, we should understand empirical causal laws in (II) as instantiations of the principle of causality in (I), where, by experiments and observations, we learn how to fill in the general template of cause–effect relation offered by the Second Analogy of Experience (I) with real grounds and their essential powers (e.g., attractive force and repulsive force). These, in turn, causally ground dispositions *D* that we see manifested in nature in a variety of

[27] A bonus of my present analysis is that it defuses the aforementioned objection about the causal inefficacy of dispositions. For it is not the case that all the causal work is done by the causal basis of properties leaving dispositions redundant. The causal basis of essential properties or powers is itself metaphysically grounded in real grounds. And it is only via the triple *ratio fiendi* modal move here described that stimulus–manifestation pairs occur and the empirical causal laws associated with them acquire nomological strength.

phenomena (e.g., balls colliding, fire burning, winds forming, and so on). That is why Kant says that empirical causal laws cannot be completely derived from the faculty of understanding (although they "stand under it") and that experience is required instead to learn the "powers of things bit by bit." Moreover, experience is required to fill in the details of how, under different stimuli conditions $Sa_1 \ldots Sa_n$, the multi-track disposition $D_{(S,M)}a$ yields different manifestations $Ma_1 \ldots Ma_n$ in the *Ratio fiendi* third modal move.

More precisely, experience enters (II) via the *Ratio fiendi* second modal move where we learn by experience and observation about these powers and conclude that there exists a real ground z for them.[28] Thus, the inference from

(I) All alterations occur in accordance with the law of the connection of cause and effect (Second Analogy).

to

(II) Event of type A (e.g., alignment of the Moon and the Earth) causes event of type B (e.g., high tide occurring – empirical causal law).

takes place by filling in the general causal template in (I) (i.e., "All alterations occur . . .") with the details of a specific power F (e.g., attractive force in this example) – and the existence of z as its real ground (i.e., the "nature" of mass) – which *causally grounds* the dispositional behavior Da_1 of massive bodies (e.g., Earth and Moon) to attract each other, as per the *Ratio fiendi* second modal move. Experience teaches us what specific kind of cause–effect connection (e.g., attraction between the Earth's mass and the Moon's mass) is in place when we observe alterations in the low and high tides at Cramond beach. Experience does not tell us that the alterations are necessary, of course. Nor does the Second Analogy tell us that either (if not in some weak, transcendental sense that does not solve the Kantian problem of inference). The necessity of the effects must then come from somewhere else.

Experience and observation enter also (III) via the *Ratio fiendi* third modal move, where by experience we learn how to track multiple effects

[28] The *Ratio fiendi* first modal move is somehow included into the *Ratio fiendi* second modal move, because its role is that of expressing the essentialist nature of these dispositional properties, which are not quiddities. As such, *Ratio fiendi* first modal move does not itself feature in the inference from (I) to (II).

under a common power F and its multi-track disposition D that we see manifested in various events M in nature. In other words, the inference from (II) to (III) "Event A_1 causes event B_1" (e.g., this particular Moon–Earth alignment today causes high tides at Cramond Island at 11:35 a.m. – instantiated empirical causal law) takes place by filling in the specific stimulus conditions $Sa_1 \ldots Sa_n$ under which the power F may find itself operating (e.g., no interference due to another massive planetary object in the Moon–Earth alignment; positions of the Sun and the Moon relative to the Earth today; position of the Moon with respect to the Earth's equator today, and so on). Once these stimuli conditions are "filled in," *necessarily* high tide at Cramond Island occurs at 11:35 a.m. today (Ma_1), as premise (III) says, via the *Ratio fiendi* third modal move. By the same token, necessarily the Earth's rotational velocity decreases over millions of years (Ma_2); necessarily coastal lines and ocean depths in both hemispheres are affected over centuries (Ma_3); and so forth. That is where the necessity of effects ultimately comes from: from the triple modal move of Kant's dispositional essentialism.

From these diverse manifestations of a multi-track disposition by simple modus ponens, whenever the right stimulus condition Sa_1 is present (premise IV), the manifestation Ma_1 *necessarily* follows (necessity of effects – conclusion V). The nomic necessity of most empirical laws, under this dispositional essentialist reading of Kant, is nothing but the expression of the natural necessity through which real grounds, and their powers qua dispositional essential properties, *causally* determine a variety of consequences in nature.

If this interpretation is on the right track, it might go some way toward explaining Kant's emphasis on the unity and systematic order of nature, according to which, once a real ground is posited for some effect (e.g., saltpeter's combustion inside the Sun), the ground of other effects (e.g., the production of winds) is posited too.

A dispositional essentialist reading of Kant on laws can solve at once the problem with the seemingly mysterious inference from (I) to (V). For it explains the necessity of the effects *in nature* via the triple modal move captured by the essentialist, the dispositional, and finally the multi-track dispositions associated with real grounds. The understanding prescribes laws to nature by providing templates of cause–effect (but also others, e.g., persistence of substance at work in the conservation of mass) that by experience and observation we learn how to "fill in" with dispositional powers, their inferred real grounds, and necessary consequences. Empirical causal laws prescribe the way nature ought to be by capturing natural necessities in the manifested behavior of real grounds qua dispositional essential properties of things, whose ultimate real essences remain unknown to us. That is why, despite the lawfulness of things in themselves being precluded to us, the lawfulness of appearances is more than just a projected or injected lawfulness. It is instead a robust lawfulness undergirded by nature's real grounds, their powers, and related modal claims.

8.4 Concluding Remarks

The main goal of this chapter was to advance an interpretation that could vindicate Kant's bold claim that the understanding prescribes laws to nature. To this end, I have elucidated the metaphysical aspect of the dispositional essentialist reading that I am defending on Kant's behalf with an eye to clarifying different kinds of necessity that Kant seems to be referring to in various passages of the lectures on metaphysics. We identified three main notions of necessity (conceptual, metaphysical, and natural necessity, respectively). They are, respectively, at work in conceptual truths, theoretical identity statements, and empirical causal laws, via three different kinds of grounds (*ratio cognoscendi*, *essendi*, and *fiendi*). This taxonomy is far from exhaustive and is meant only to map out (tentatively) the territory of lawlike claims and their necessity in Kant.

The nomic necessity at play in the majority of empirical laws of nature, I have claimed, is *natural necessity* captured by real ground–consequence relations. I have clarified three modal moves that are involved in the way in which real grounds qua *rationes fiendi* bring about their effects in the harmonious and essential order of nature, an order that is ultimately subsumed under God, yet nomically independent of Him, to Kant's eyes.[29]

The metaphysical quandary of the problem of inference evaporates at once. The faculty of understanding has a purchase on nature because it provides general templates under which stand many empirical laws. And, crucially for my story, these laws capture the natural necessities with which dispositional essentialist properties bring about their effects.

[29] "The forces of nature and the causal laws which govern them, contain the ground of an order of nature. This order of nature, in so far as it embraces a complex harmony in a necessary unity, has the effect of turning the combination of much perfection in *one* ground into a law. Thus, different natural effects are, in respect of their beauty and usefulness, to be regarded as subsumed under the essential order of nature, and, by that means, as subsumed under God" (*OPA* 2:107). On this point, see Massimi 2014.

Thus, nomic necessity qua natural necessity supervenes on nature's essential powers and their dispositional behavior. While the underpinning metaphysics of my reading of Kant on laws is dispositional essentialist, it should also be clear that for us to *cognize the necessity* of the laws, that is, to stumble into empirical regularities and identify them as modally robust (as we expect laws of nature to be), they *ought to stand under the laws of the understanding* and "fill in" the general templates provided by the System of the Principles.

Kant knew that Humean regularities are where one wants to find them. What makes some of them laws of nature (while others are not) is precisely the fact that some of them are undergirded by grounds and their necessary effects, while accidentally true ones are not.[30] The *Ratio fiendi* triple modal move presented in Section 8.3 shows how to move from general principles of the understanding to instantiated empirical laws, whose necessity is not "injected" or downstream to the necessity of the principles themselves. It is instead grounded in the natures of the things and their dispositional essentialist powers, which we come to know "bit by bit" by experience.[31]

Turning to the epistemological quandary, we *can cognize* the *necessity* of empirical causal laws (which is derived neither a priori from the understanding nor a posteriori by experience), because these laws "stand under" the formal template of causality, under which only it is possible for us to carve nature's empirical manifold according to modally robust regularities. The quandary was based on erroneously conflating the necessity (qua Postulate of Empirical Thinking) that attaches to the principle of causality, with the nomic necessity that – under my interpretive reading – Kant saw as natural necessity under a dispositional essentialist view of real grounds. The former provides only necessary rules for ordering our representations according to cause–effect. The latter is evidenced by the plurality of effects traceable to the same real ground in nature.[32]

[30] "But there are cases where something is posited, and another thing is posited after, yet where the one is not a ground of the other. E.g., when the stork comes, good weather follows. But to posit [*ponere*] does not mean something follows the other accidentally; for the stork could also be brought on the mail coach" (28:549).

[31] Hence my metaphysical picture of powers and natural necessity differs in significant ways from Friedman's reading of causal laws whose necessity is injected from the necessity of the Second Analogy and, more generally, of the principles of the understanding as constitutive a priori for our experience of nature (see Friedman 2014 and Chapter 10 in this volume).

[32] My account of natural necessity would then be compatible with readings of Kant that place emphasis on the principle of systematicity (esp. in the third *Critique*) as a way of understanding

What ultimately allows us to encounter nature as ordered, lawful, and harmonious is our ability to cognize (neither the essences nor the nature of things but) the multifarious manifestations of the powers of things "bit by bit." The faculty of understanding, I suggest, offers nomological cookie-cutters with which we learn how to carve out the empirical manifold at its joints, namely along the lines of its natural necessities. In this sense, and in this sense only, I conclude, can Kant successfully (and nontrivially) claim that our understanding prescribes laws to nature, a nature that never ends to manifest itself as ordered, lawful, and harmonious in its seasonal and daily regularities.

Acknowledgments

I am very grateful to Karl Ameriks, Angela Breitenbach, Andrew Stephenson, and Eric Watkins for reading earlier drafts and providing very detailed and constructive comments. I owe special thanks to Bryan Pickel for helpful discussions on Section 8.3 and the metaphysics of grounding. Lorenzo Casini, Andrew Chignell, Fabrice Correia, Steven French, Paul Guyer, James Messina, Alissa Ney, and Marcel Weber provided helpful feedback on various points. I am very grateful to James Collin for his help with proofreading the chapter. Needless to say, any errors remain my own. Finally, I gratefully acknowledge the support of the Leverhulme Trust for the international network IN-081 "Kant and the Laws of Nature," of which I had the honor of being the PI (2012–2015) and from which this research originates.

why particular regularities count as laws (insofar as they are part of a system of empirical laws). Yet I part ways with these readings because I do not see the nomic necessity of the empirical laws as *itself* a consequence of being part of a system of laws (for reasons that I discuss elsewhere; see Massimi, forthcoming). Thus, I share with these readings of systematicity the idea that *lawlikeness* (if understood as an epistemic feature of *cognizing* some regularities as laws) might well be a function of them being part of a system of laws. E.g., *our ability to cognize* the attraction between the Earth and the Moon as *lawlike* in causing tides has to do with us having a system of Newtonian mechanics in place; in another world *w′* where no such system were in place, the same phenomenon could not be cognized as lawlike. But I diverge from these readings of systematicity in contending that the *lawfulness* (understood as a metaphysical feature of why some regularities are indeed laws of nature) has nothing to do with them being part of a system of laws. According to the view defended in this chapter, in another world *w′* where Newton's system were not in place, the attraction between the Earth and the Moon would still be lawful for Kant, because it would still be the expression of the natural necessity with which once gravity is posited, tidal effects necessarily occur.

CHAPTER 9

Kant on Mathematical Force Laws

Daniel Warren

In the *Metaphysical Foundations of Natural Science*, Kant is clearly committed to the idea that the laws of the fundamental forces of matter can be known only empirically.[1] In the General Remark to the Dynamics chapter of that work, he says:

> [I]f the material itself is transformed into fundamental forces (whose laws we cannot determine a priori, and are even less capable of enumerating reliably a manifold of such forces sufficient for explaining the specific variety of matter), we lack all means for *constructing* this concept of matter, and presenting what we thought universally as possible in intuition. (*MF* 4:525)

And then a bit later in the same section he writes:

> [N]o law of either attractive or repulsive force may be risked on a priori conjectures. Rather, everything, even universal attraction as the cause of weight, must be inferred, together with its laws, from data of experience. (*MF* 4:534)

And yet Kant also presents what *seem* like a priori arguments for the inverse-square character of the law of gravitational attraction. Such passages appear in the *Prolegomena* (§38), in the pre-Critical *Physical Monadology*,[2] and, again, in the Dynamics chapter of the *MF*.[3] These arguments, which are, in part, explicitly geometrical in character, concern the properties of spheres, specifically the fact that the area of a sphere varies with the square of the radius. Moreover, there are closely parallel

[1] See also A207/B252, near the end of the Second Analogy: "Now how in general anything can be altered, how it is possible that upon a state in one point of time an opposite one could follow in the next – of these we have *a priori* not the least concept. For this acquaintance with actual forces is required, which can only be given empirically, e.g., acquaintance with moving forces, or, what comes to the same thing, with certain successive appearances (as motions) which indicate such forces."

[2] In the second paragraph of the Scholium to Proposition X (1:484)

[3] In the first Remark attached to Proposition 8, paragraphs labeled (2) and (4) (4:519–21)

arguments presented in the same sections of the *MF*,[4] as well as the *Physical Monadology*,[5] purporting to establish some kind of inverse-cube law for matter's original repulsive force, and to do so by appeal to the fact that the volume of a sphere[6] varies with the cube of the radius.

In this chapter I want to address two questions: what is the character of the seemingly a priori arguments for these fundamental force laws? And do these aprioristic considerations leave any room for the "data of experience" in grounding these laws? I begin with the first question, and in particular, I am concerned with the relative roles of mathematical and more metaphysical (perhaps "transcendental") aprioristic lines of thought in Kant's treatment of these force laws.

9.1 Metaphysical Tasks and Mathematical Tasks

I will focus on the account given in the *MF* Dynamics chapter. That text gives us especially helpful material to work with for evaluating the status of the mathematical considerations in Kant's views about the essential forces of matter. The first thing to note is that, although these arguments for the specific mathematical form of the basic force laws are part of the *text* of the *MF*, they occupy a *marginal* position. What I mean is this. The statement of the inverse-square law and of the inverse-cube law and the arguments for them are not found in any of the Explications or Propositions of the Dynamics chapter of the *MF*, nor in any of the proofs of these Propositions[7] nor of any other chapter of the *MF*, for that matter. These laws and the considerations in support of them are presented only in various Notes and Remarks appended to, and going well beyond the scope of, the last of the Propositions in the Dynamics chapter. That seems very much in line with Kant's conception of the *MF*. For he says that the establishment of these laws falls outside the essential tasks of that work. He does not want it "to be viewed as necessarily belonging to the goals of my metaphysical

[4] In the first Remark attached to Proposition 8, paragraphs labeled (1), (3) and (4) (4:519–21)

[5] In the first paragraph of the Scholium to Proposition X (*PM* 1:484)

[6] In the *MF*, the sphere is regarded as having an infinitesimal radius; in the pre-Critical *Physical Monadology* it has a finitely large radius.

[7] What Kant *does* argue for here is not the specific mathematical form of the force laws; what he argues for is a more general qualitative feature of the dependence of the force on distance. In Proposition 3 of the Dynamics chapter, he argues that the expansive force due to the mutual repulsion of the parts of matter must become infinitely great in the limit as the volume of the matter is compressed to zero. And in Proposition 8, Kant argues that any two pieces of matter must attract one another to some degree not matter how distant they are. The attractive force diminishes as the distance increases, but never goes to zero.

treatment of matter" (*MF* 4:522–523). And although the determination of these laws and their relation is needed in order to construct what he calls "the dynamical concept of matter, as that of the movable filling its space (to a determinate degree),"[8] this is not a task Kant thinks metaphysics is sufficient for carrying out, and the task is considered by him, at least in part, to be:

> a purely mathematical task, which no longer belongs to metaphysics – nor is metaphysics responsible if the attempt to construct the concept of matter in this way should perhaps not succeed. For it is responsible only for the correctness of the elements of the construction granted to our rational cognition, not for the insufficiency and limits of our reason in carrying it out. (*MF* 4:517–18)[9]

In this passage, Kant is pointing out that the resources of metaphysics need to be supplemented *not* by the data of experience, but rather by mathematics, with all of the differences in method and in the prospects of completeness that Kant believes the contrast between mathematics and metaphysics entails. Moreover, we can see from this that when we ask whether, or to what extent, Kant thinks the basic force laws can be known a priori, this question needs to be distinguished from the more limited question about their *metaphysical* foundations, which Kant believes to be essential to the task of the *MF*. For there are a priori (mathematical) considerations involved in establishing the force laws and their relation to one another that go beyond the scope of metaphysics.

[8] Kant writes here: "But for this [construction] one needs a law of the ratio of both original attraction and repulsion at various distances of matter and its parts from one another" (*MF* 4:517). I take it that what Kant is especially concerned with here is whether the mathematical form of the force laws will be such that they always admit of a stable balancing point, a distance below which repulsion is greater than attraction and above which attractive force predominates. This is already seen as a central concern of Kant's theory of matter in the pre-Critical *Physical Monadology* (1:485).

[9] Later, after presenting the arguments for the specific mathematical form of the laws of attraction and repulsion, Kant writes, "I declare, furthermore, that I do not want the present exposition of the law of an original repulsion to be viewed as necessarily belonging to the goals of my metaphysical treatment of matter. Nor do I want this latter (for which it is enough to have presented the filling of space as a dynamical property of matter) to be mixed up with the conflicts and doubts that could afflict the former" (*MF* 4:522–523). The determination of the specific mathematical form of neither the attractive nor the repulsive force belongs to the goals of Kant's "metaphysical treatment of matter." But that is not the reason he speaks of "conflicts and doubts" here. The worries Kant is expressing concern his treatment of *repulsive* force in particular, possibly on account of its appeal to the idea of an infinitely small distance between parts of matter. And Kant never expresses corresponding worries about his treatment of the inverse-square character of attractive force. The reason these arguments do not belong to Kant's "metaphysical treatment of matter" is that they are in part geometrical in character, not because they are necessarily afflicted with "conflicts and doubts."

9.2 The Universal Law of Dynamics and the Diffusion of Forces

The arguments Kant provides for the inverse-square law of attraction and for the inverse-cube law of repulsion crucially depend on the following principle:

> [T]he action of the moving force, exerted by a point on every other point external to it, stands in inverse ratio to the space into which the same quantum of moving force would need to have diffused [*sich ausbreiten*], in order to act immediately on this point at the determinate distance. (*MF* 4:522)

Kant calls this "the universal law of dynamics" (*MF* 4:522). That name might lead a reader to expect that this law will occupy a prominent place in the presentation of the Dynamics chapter. But it does not, at least not as an explicitly enunciated claim. Nor does Kant give us any sense here of what kind of grounds might be put forward to support it. The only place it is stated is in the Remarks attached to the last Proposition of the Dynamics chapter, and there, only in connection with the justification of the inverse-square and the inverse-cube laws. Nevertheless, it is a principle that is extremely important to the way Kant conceives the contribution of the Dynamics chapter to the central aims of the *MF* as a whole, which concern the possibility of mathematizing the doctrine of nature. I said that the universal law of dynamics is not stated explicitly in the main thread of the Dynamics chapter's Explications and Propositions (including the proofs of the latter). It is, however, essential to these arguments at two crucial points, although more qualitatively, rather than by way of supporting a determinate mathematical force law: first in the argument that matter cannot be penetrated by a force of compression, no matter how strong (Proposition 3);[10] and second in the argument that attractive force is exercised on all bodies no matter how far away they may be (Proposition 8).[11]

In his statement of what he calls the universal law of dynamics and in its applications, Kant talks about the "diffusion" (*Ausbreitung*)[12]

[10] The implicit, but clear, appeal to this principle is found in a Remark appended to the proof of Proposition 3. This is a Remark that merely serves to explain the preceding proof, in contrast to the Remarks and Notes following Proposition 8, where Kant's intention is to go well beyond the scope of that Proposition and its proof.

[11] The role of this principle in the proof of Proposition 8 will be discussed in more depth in what follows.

[12] In these passages from the Dynamics chapter, Kant seems to use interchangeably both the noun forms "*Verbreitung*" and "*Ausbreitung*," as well as both the verb forms "*(sich) verbreiten*" and "*(sich) ausbreiten*."

of the force or of the quantum of force in a space.[13] I want to make two points about this talk of diffusion here. First, what is diffused is *not* some kind of substance, some kind of material or fluid, though it seems natural to picture it that way. What Kant says is diffused is the *effect* or *action*[14] (*Wirkung*) of the force or, sometimes, its quantity. It is hard to work out a precise positive characterization of the ontological status of what diffuses,[15] but I do not think that question needs to be addressed for my purposes here. Second, diffusion is not necessarily to be regarded as a temporally extended process. It is not that, in the course of diffusing, the activity of a force shifts or extends from one place to another. The representation of the diffusion is just the representation of the spatial distribution of something's activity, which Kant also calls its sphere of influence. When Kant speaks of a body's attractive force diffusing through all space, he does not mean that its activity is first closer to the body and then later, further away. The diffusion through a space of a body's essential repulsive forces is its "filling" that space, but this diffusion is no more a temporal process than filling the space is. On the other hand, this does not mean that the distribution of the activity of a force is unchanging. If an attracting body changes its shape from a sphere to a flattened ellipsoid, this will be accompanied by a change in the distribution of its activity in space. The point is that diffusion is not itself a *temporal process*, not that the diffusion of activity can never *change*.

9.3 Kant's Account of Intensive Magnitudes

So, I have said that Kant enunciates a claim he calls the universal law of dynamics, which is that a force or its activity, or more properly, its quantity, is inversely proportional to the space into which it must diffuse in order to act immediately at a given point. But what is behind this claim? What, in Kant's view, grounds it? I want to suggest that what grounds it is the picture Kant has of what he calls intensive magnitudes.

The two essential forces of matter discussed in the Dynamics chapter, attraction and repulsion, can be said to have intensive magnitude.[16]

[13] Kant also uses the same language in his argument about the inverse-square law of attraction in section 38 of the *Prolegomena* (4:321) when he says that the attractive force spreads itself (*sich verbreitet*) into a space.

[14] In context, "*Wirkung*" is often translated as the "action" of the force.

[15] For it is often only potential or possible effects, rather than actual effects, that are said to diffuse.

[16] I discuss Kant's views about intensive magnitude in chapter 1 of Warren 2001.

They can be regarded as being of greater or lesser strength, even if these forces are represented as acting from a mere point in space. The concept of intensive magnitude is understood in contrast to that of extensive magnitude. When we represent a quantity of something as extensive, we represent it as made up of mutually *external* smaller parts, for example, a certain quantity of space, or of matter, insofar as it is represented as composed of smaller parts outside one another. An intensive quantity, on the other hand, is a degree or an intensity, for example, of heat, of brightness, or of speed. When we represent a quantity of something as intensive, we do not represent it as made up of mutually external smaller parts. In a Reflection written between 1788 and 1790, Kant says,

> One cannot say, e.g., that heat consists of lukewarmths [*Lauigkeiten*], one therefore determines its magnitude not according to the parts which it contains . . . (*Refl.* 5663, 18:322, my translation)

I will come back to this Reflection shortly, and I will quote it at greater length.

Here I will just note that this point about *warmth*, as an intensive magnitude, is meant to hold for intensive magnitudes generally. An intensive magnitude is not represented as made up of distinguishable parts, each with a smaller intensive magnitude.

Now, because of this feature of intensive magnitudes, there arises a prima facie problem in representing intensive magnitudes as quantities in the full sense. By that I mean specifically, that, prima facie, we lack a means of representing what we understand by the addition or subtraction of such intensive magnitudes. For extensive magnitudes, this is not a problem. The sum of two extensive magnitudes of the same sort is just an extensive magnitude with these two magnitudes as its *parts*. But with intensive magnitudes we cannot take for granted the idea of distinguishable smaller parts that make up the whole magnitude. I am taking it (and this will be important in the next section) that for Kant, homogeneous intensive magnitudes, as such, *do* admit of comparison as greater or less, or as equal, in their degrees of intensity. But that is not sufficient for the representation of *addition or subtraction* of such quantities. One can know that intensity **a** is greater than intensity **b**, without knowing *how much* greater **a** is than **b**, that is, without being able to represent that intensity **c**, such that **a** = **b** + **c**. But because we do not represent intensive magnitudes as composed of smaller parts, the most straightforward way of making out the idea of addition and subtraction is lacking. How then are we to represent intensive magnitudes as quantities in the fuller sense? Here I want to quote at greater length the Reflection I cited earlier:

> One cannot say, e.g., that heat consists of lukewarmths [*Lauigkeiten*], one therefore determines its magnitude not according to the parts which it contains, but rather according to the effects which it produces, e.g., that it expands a body. And one can thereby ascribe to it not a genuine magnitude, but rather a degree. (*Refl.* 5663, 18:322, my translation)

One must not be misled by Kant's saying that heat is not a "genuine" magnitude; it is clear that in this context he intends to restrict the qualification "genuine" to cases of *extensive* magnitude. What is certain is that Kant repeatedly and in many places acknowledges "degree" as a type of magnitude (intensive magnitude) and that in this very passage, the point is that we can determine the magnitude of something, even though it is not made up of smaller parts external to one another, as an extensive magnitude is. We do so by reference to the effects it produces, in this case, by reference to the fact that it expands a body, that is, by means of some kind of thermometer (perhaps in which it is a volume of mercury that is expanded). This will enable us to say not just that a given intensity of heat is *greater than* another, but *how much* greater it is.

This is not just an offhand observation about how we might measure intensive magnitudes. It is part of a view about the sense in which intensive magnitudes *are* magnitudes. Kant thinks that the representation of addition is, at least in part, constitutive of the determinate concept of a quantity. He writes in the *MF* that "[t]he determinate concept of a quantity is the concept of the generation of an object through the composition of the homogeneous" (4:489). This is in a context in which Kant is explaining the composition of motion as "the representation of the motion as the same as [*einerlei mit*] two or more motions combined together" (ibid.), which as the text that immediately follows makes clear, is a matter of representing the *addition* of motions. So unless we succeed in representing the addition of quantities of a given sort, we will not have managed to bring this sort of thing under the "determinate concept of a quantity." And in the *Critique*, when Kant is asking how we can regard the various categories as having content, and he comes to the category of quantity, the corresponding question is "how a thing can be one with [*einerlei mit*] a number of others taken together" (B288), a formulation that is very much the same as that found in the *MF* passage just discussed.

Moreover, Kant's account of intensive magnitude as the magnitude of a ground is one that is repeated many times in different *Reflexionen* and in different *Lectures on Metaphysics* (cp. *Refl.* 4411, 17:536; *Refl.* 6399, 18:705; 28:425, 507, 637). In those places, intensive magnitude is described as the magnitude of a thing as a ground, in contrast with extensive magnitude, which is said to be the magnitude of a thing as an aggregate.[17] Here are examples of such passages (my translations):

> The magnitude of a thing as aggregate is extensive, as ground, intensive. (18:705)

> Intensive magnitude is that of a ground, extensive, [that] of an aggregate. (28:425)

> The magnitude of the quantum considered as an aggregate is the extensive. But the magnitude of a quantum [considered] as a ground is the intensive. (28:637)

Kant's view is that even if an intensive magnitude is not in the usual sense made up of parts, it can be the *ground* of a plurality of consequences. This view is seen in the following passage:

> Now that which in perception contains no manifold is called an intensive magnitude. In it we distinguish nothing manifold; rather, *the manifold is here represented only in its consequence* and therefore, it is here the magnitude of a ground. (28:507, my translation and added emphasis)

When we consider a magnitude as ground, then the aggregate of its consequences (really what Kant probably means here is the aggregate of its *possible* consequences) can be thought of as made up of parts that are external to one another, typically, spatially external to one another. So an intensive magnitude can be thought of as a quantity in the full sense, a

[17] Sometimes this relation of ground and consequence is to be understood as that of cause and effect. But in some contexts it is to be construed in a much more abstract sense. It was noted earlier that Kant considers speed to be an intensive magnitude. This view figures centrally in the Phoronomy chapter of the *MF* (4:493–494). And yet Kant emphasizes the requirement that, in that chapter, any appeal to causality be avoided. So if, in this context, "the quantity of speed, considered as quantity, we regard not as aggregate, but as ground" (28:637, my translation), he must be understanding speed as "ground" in a very broad sense. Here I merely wish to note that Kant's strategy in the Phoronomy chapter is to explain the addition of (instantaneous) speeds in terms of the addition of corresponding motions, which are temporally extended processes (*MF* 4:494). Kant has already shown us what it means to add motions, each represented in a distinct reference frame. And it is in these terms that we are, in turn, to understand the addition of the corresponding speeds. Perhaps Kant here treats the representation of the motion (in a given reference frame) as the basis on which we can have knowledge of the speed, and then relies on a traditional formulation, according to which, this appeal to a "*ratio cognoscendi*" is identified as a case of knowing the ground by means of its consequences (*NE* 1:392–393).

sense for which addition and subtraction are defined, but – at least according to Kant – only by reference to the aggregate of its consequences, because that *can* be regarded as an extensive magnitude.

9.4 Intensive Magnitude in Photometry: Diffusion of Effects

Note that this is not to say that the *consequences* do not also have *intensive* magnitudes as well. And if they do, we might worry that we are, for that reason, right back where we started. That is, we might worry that unless we can analyze intensive magnitudes in terms of quantities that are *purely* extensive magnitudes, like regions of pure space, which have no intensive magnitude, then we will have made no progress. But that is not so. For we assumed earlier that we were already able to compare intensive magnitudes as greater than, less than, or equal to each other. Thus, if we find homogeneous consequences that have the *same* intensive magnitude, then we can compare them quantitatively by just comparing their extensive magnitudes. The paradigmatic cases derive from photometry. For example, according to *Metaphysik Volckmann*, Kant says that

> the illuminating power of a wax light [*Wachslicht*] is intensively greater than that of a tallow light [*Talglicht*], for with the first we would be able to read at a distance of 2 feet, but with the latter only at a distance of 1 foot. (28:424–425)

The intensive magnitude of the effect, that is, the degree of illumination at the threshold needed for reading, is the *same* in both cases. And this is what allows for a comparison of the intensity of the causes, that is, the flames, in the two cases. How does this comparison work? We draw on the fact that the distance in the one case is double that in the other. But then we need to go beyond that and consider the aggregate of all the possible consequences, say, of the *wax* flame, that have the same intensity as the consequence we were given: namely, the minimum degree of illumination needed for reading. And then, largely on the basis of symmetry considerations, we say that the effects will be the same at any point that is two feet from the wax flame, that is, over the whole spherical surface with a two-foot radius and centered at the flame. In such a case, when the spherical surface is evenly illuminated, we can say, for example, that the quantity of the consequences over an arbitrary two-square-foot area of that surface will be twice the quantity of the consequences over an arbitrary one-square-foot portion of the same surface. Now, once again comparing the wax flame and the tallow flame, we can see that the former can illuminate a spherical surface with a two-foot diameter with the same intensive magnitude as the latter illuminates the surface of a one-foot-diameter sphere. It is based on this reasoning, in which we compare spaces, here spherical surfaces that have an *equal* intensive magnitude of illumination, that we can say that the ground of this illumination, the flame, has four times the intensive magnitude in the one case as it does in the other. Again, the problem is to compare *grounds* with different intensive magnitudes. And the point is that by comparing *effects* with *equal* intensive magnitude, we reduce the problem to one of comparing extensive magnitudes.

That's not quite the end of the story. So far we have compared the two-foot-radius sphere around the wax candle with the equally bright one-foot-radius sphere around the tallow candle. What about the problem of comparing the intensity of illumination on spheres of different radii but with the *same* center, say, where the wax flame is located? That's the kind of thing that we are *really* interested in here, insofar as our focus will be on the analogy with the universal law of dynamics. So, the wax flame has an intensity such that it can evenly illuminate a spherical surface surrounding it to a given degree. When we compared this with what other kinds of flame can do by way of illumination, we took this effect (the given degree of illumination), multiplied by the total area of the surface (the two-dimensional space that is being illuminated), to represent the intensity of the ground, that is, the flame. That is, the intensity of the ground (the flame) is represented by the intensity of the consequence, in other words, the degree of illumination, times the total area illuminated at this degree. Now here is the crucial move. It involves some presuppositions, which I'll try to bring out shortly. The choice of the radius for the sphere being illuminated was arbitrary. If the intensity of illumination times the area of the sphere gave us a good measure of the intensity of the flame, then we should get the same value no matter what sphere centered on that flame we examine for intensity and area. And if that were true, then we would have gotten the inverse relation between the degree of illumination and the amount of space (here, a two-dimensional space, i.e., a spherical surface) in which the effect is diffused. The degree of illumination, in other words, would have to vary inversely with the square of the distance from the flame.

But now it's natural to want to object to all this. Something must have gone wrong here. You can't get laws of photometry just from this apparently aprioristic reasoning. For example, what if the medium absorbs light to some degree? What if the candle is in a smoky room? Then the product of the degree of illumination and the area of the spherical surface with a larger radius will *not* be equal to that product taken on a smaller sphere. It will be less. And, by the same token, the method suggested by Kant for comparing the intensity of the tallow and the wax flames would not work either if the candles were located in a smoky room. But, now, returning again to the case of different spherical surfaces around a *given* flame, I think it is natural to say that the reason the product is less on the larger sphere is that the total effect of the flame is decreased by its interaction with the smoky medium. That is, it seems natural to analyze the discrepancy in causal terms. The medium dissipates the total illuminating effect of the flame. The question Kant would focus on is this: wouldn't the inverse relation between the intensity and the area at least hold in the case where there was no dissipating medium? This kind of question will probably seem to *us* a wholly empirical question. But I think that, properly understood, for Kant it was not, or at least not wholly.

9.5 Intensive Magnitude in Dynamics: Diffusion of Effects

Let's shift our attention to the case of attractive force acting immediately, that is, without any causal mediation, at a distance. I take it that this kind of dynamical force, conceived as a causal power belonging to the attracting body, is treated by Kant as an intensive magnitude. And if that is so, then its magnitude is the magnitude of a ground. That is, we can compare these magnitudes by considering the extensive magnitude of its consequences.[18] More specifically, the intensity of two grounds can be compared, by comparing the extensive magnitudes of effects with the same intensive magnitude. But then the objection gets raised, as in the case of photometry, of how do we know that there is not, for example, some dissipation of the action of the attracting body, for example, as we get further away from

[18] This is not to say that we can compare intensive magnitudes of the grounds *only* by comparing the extensive magnitudes of effects with the same intensity. E.g., the ratio of the attractive power of two planets, Jupiter and Saturn, may be estimated by comparing the period of a moon revolving around the one (at a certain distance from it) with that of a moon revolving around the other planet (at the *same* distance from it). Here, rather than the intensive magnitude of the effect, it is the distance between the planet and its moon that is the same in both cases. Kant discusses such a case in the *MF* Dynamics chapter (4:515). There, the intensities of the effects produced by the attraction of the two planets can be compared by comparing the periods of their moons. That the distance is the same in both cases is what, according to Kant, allows him in this case to identify the ratio of the intensities of the effects with the ratio of the attractive powers and thus of the masses of the attracting planets. (Cp. 4:516, at the end of the Note to Explication 7.) But Kant's idea in claiming that the intensive magnitude of something is its magnitude *as ground* is that the special method of comparing effects at the same distance is to be integrated into a general account according to which the plurality of effects – effects of the same intensity – gives us the fundamental idea of the "total quantum" of the effects, which can then allow us to compare the intensity of the effects at *different* distances, in accordance with the universal law of dynamics.

it? This is where Kant will appeal to the fact, which he takes that he has established, that attraction acts *immediately* at a distance. And because it acts immediately at a distance, he says, "no intervening matter sets limits to the action of an attractive force" (*MF* 4:516).

Kant thinks that it doesn't make any difference if there is matter of some sort between the attracting body and the place it acts; it will act just as it would if the space in between were absolutely empty. This point is repeated just a bit later, and then further consequences are drawn, when Kant presents his proof of Proposition 8, the claim that attractive force extends to infinity:

> Because the original attractive force belongs to the essence of matter, it also pertains to every part of matter to act immediately at a distance as well. But suppose there were a distance beyond which it did not extend. Then this *limiting* of the sphere of its activity would rest either on the *matter* lying within this sphere, or merely on the magnitude of the *space* in which it diffuses this influence. The first [case] does not hold; for this attraction is a penetrating force and acts *immediately* at a distance through that space, as an empty space, regardless of any matter lying in between. The second [case] likewise does not hold; for, since every attraction is a moving force having a degree, below which ever smaller degrees can always be thought to infinity, a greater distance would indeed be a reason for the degree of attraction to diminish in inverse ratio, in accordance with the measure of the diffusion of this force, but never for it to cease altogether. Thus, since there is nothing that has anywhere limited the sphere of activity of the original attraction of every part of matter, it extends beyond all specified limits to every other matter, and thus throughout the universe to infinity. (*MF* 4:516–517)

So, assume that a body endowed with a causal power (an attractive force) produces effects, which are limited in their intensive magnitude as we get further from that body. Kant's point is that this diminution of the body's action can have two very different kinds of ground. It could be due to the intervening matter. And I take it that, in that case, the diminution has some kind of causal basis; it is something that can be traced to causal powers and susceptibilities of the intervening matter. But in the case of a force that acts immediately at a distance, the intervening matter is, so to speak, circumvented, along with any possible role it could play in diminishing the effects the body can exert. The second kind of ground for the diminution of the effects produced by a body, according to Kant, is just the magnitude of the space itself. But this diminution by the magnitude of space, unlike a diminution due to some intervening matter, is not itself a causal process. Empty space, Kant assumes, is *causally inert*. It neither produces effects nor is it affected by anything. And I think that this is crucial to Kant's claim that "a greater distance would indeed be a reason for the degree of attraction to diminish in inverse ratio, in accordance with the measure of the diffusion of this force." The magnitude of the space can be a reason for the effects to be more spread out. It is a reason for a change in the spatial distribution of the effects. But, as I understand the picture here, if all that happens is a redistribution in space of the effects, the idea is that the *total* magnitude of the effects is unchanged.[19] What Kant calls the "universal law of dynamics," the law that the intensity of the effect is in inverse proportion to the space in which the force diffuses, derives from his conception of the contrast between a space that is filled with matter and a space that is empty. The former can bring about a diminution of the total effect, but the latter cannot: it can bring about only a change in the spatial distribution of that effect, that is, a change in its extensive magnitude and, with that, a corresponding change in the intensity of that effect. Kant writes:

> Of any force that acts immediately at various distances, *and is limited*, as to the degree with which it exerts moving force on any given point at a certain distance, *only by the magnitude of the space into which it must diffuse* so as to act on this point, one can say that in all the spaces, large or small, into which it diffuses, it always constitutes an equal quantum, but [also] that the degree of its action on that point in this space is always in inverse ratio to the space, into which it has had to diffuse, so that it could act on this point. (*MF* 4:518–519, emphasis added)

A force "limited . . . only by the magnitude of the space into which it must diffuse" always constitutes an "equal quantum" because it is only by causal interactions that this "total quantum" can be diminished (or increased), and because a quantity of space is by itself endowed with no causal powers, and so, none that could result in such a diminution (or increase).[20] I take it that Kant would not consider this to be an empirical point. But again, I think that there are presuppositions here that will need to be brought out.

[19] Here the "total" magnitude of the effects is conceived as the product of the intensive magnitude of the effects and the extensive magnitude of the space (here, a two-dimensional space), focusing only on the space where effects having equal intensive magnitudes are found. This way of characterizing the "total" quantum of the effects is adequate for the kinds of forces and spaces that Kant considers. It is a much trickier problem to characterize it in a way that would be adequate in a completely general way.

[20] I take it that this, by itself, is not to say that this "total quantum" is conserved over time. The "total quantum" of the effects is equal at different distances, *assuming* the degree of the causal power (their ground) remains the same. If that degree were to change, then presumably the "total quantum" would change as well.

Kant is not claiming that the intensity of the effects due to *every* kind of force or causal power, or even every moving force, that a body might have, decreases in inverse proportion to the quantity of space through which it has to diffuse. In fact, the original statement of the universal law of dynamics at 4:522 explicitly restricts its application to the *immediate* effects of the force. A simple example of an action that is not immediate would be the propagation of a motion from one place to another by an intervening medium that is elastic. If the effects are produced by some mechanism, a specific chain of causes and effects, that is, if the action of the force is mediated, then the universal law of dynamics does not apply. Kant thinks that, in the case of the essential attractive force of matter, he has *established* the immediacy of its action at a distance (in Proposition 7 of the Dynamics chapter). And although we may characterize other kinds of causal power in terms of the effects it produces, we cannot in general assume that these effects are the *immediate* results of the power. So we cannot, in general, compare the intensities of homogeneous causal powers simply by comparing the circumstances in which they produce effects of equal intensity, as we did in the example of the wax and tallow candles. Where the immediacy of action is not assumed, we will be able to rely on the intensity of the effects in order to estimate the intensity of the causal power or force responsible for them, but only in conjunction with an appeal to a lot of empirical regularities about the causal mechanisms by which these effects are produced.[21]

There is another kind of restriction on applying the universal law of dynamics, though, which seems to me conceptually distinct from the immediacy restriction. And this is not made explicit in the statement of that law. This concerns the exclusion of *derivative* forces from the set of forces to which this law is supposed to be applicable. I take it that a force is said to be *derivative* if its effects can be reduced to the effects of more basic forces.[22] That the universal law of dynamics is not meant to apply to derivative forces comes out clearly in a Remark on the proof (of Proposition 3) earlier in the Dynamics chapter, where Kant attempts to show that

[21] E.g., we may need to appeal to the empirically known properties of a medium that intervenes between the cause and the effect.

[22] An action of a certain sort could, in principle be immediate, but derivative, if it is the sum of the effects of several distinct fundamental causal powers acting immediately at some point. That is, this action can be reduced to the actions of more fundamental powers, none of which, however, work through some mediating causal chain.

> Matter can be *compressed* to infinity, but can *never* be *penetrated* by a matter, no matter how great the compressing force of the latter may be. (*MF* 4:501)

In the Remark associated with this Proposition, Kant says:

> In this proof I have assumed from the very beginning that an expanding force must counteract all the more strongly, the more it is driven into a smaller space. But *this would not in fact hold for every kind of merely derivative elastic forces*. However, it can be postulated in matter, insofar as essential elasticity belongs to it, as matter in general filling a space. For expansive force, exerted from every point, and in every direction, actually constitutes this concept. But the same quantum of extending forces, when brought into a smaller space, must repel all the more strongly at every point, the smaller the space in which this quantum diffuses *[verbreitet]* its activity. (*MF* 4:501, emphasis added)

Kant speaks of an expanding force here, and this is just the basic repulsive forces exerted between the *parts* of a given piece of matter, which tend to make it expand or to resist an externally applied force of compression. Here, the (extensive) magnitude of the space in which the force diffuses is a bounded three-dimensional space rather than the two-dimensional spherical surface we considered in the case of attractive force. Again, Kant talks about a given "quantum" of force, which can, as he puts it, "diffuse" in a larger or smaller space. Here, with compression, as the three-dimensional space, that is, as the extensive magnitude, decreases, there is a corresponding increase in the intensity of the effects, without any upper limit. This is what would be required by the universal law of dynamics. But Kant points out that this does not necessarily occur if the force in question is a derivative force. The effects of a *derivative* force may reach a maximum limit in their intensity, or they may even decrease with an increase in compression.

So, again, one cannot, in general, infer the intensity of the ground, the causal power, from the intensity of the effects, simply by using the universal law of dynamics. And again, as we said a bit earlier, in most cases we will be able rely on the intensity of the effects, in order to estimate the intensity of the causal power or force responsible for them, only in conjunction with an appeal to a lot of empirical regularities about the causal mechanisms by which these effects are produced. However, Kant's point, as I understand it, is that, where the causal powers are sufficiently basic – where they are neither derivative nor their effects causally mediated – then some consistently applicable notion of the "total quantum" of effects will play an especially fundamental role in our conception of the intensity of the ground, that is, the causal power. And in that case, we will be able to rely on the comparison of the extensive magnitudes of the effects in order to compare the intensity of the grounds by appealing to the universal law of dynamics.

9.6 A Priori Aspects of Force Laws and the Room They Leave for the "Data of Experience"

In the *MF*'s Dynamics chapter, in the passages where he grounds the inverse-square law of attraction and the inverse-cube law of repulsion, Kant relies on several distinct considerations. First, he relies on the claims that any force that diffuses evenly within a surface will decrease inversely with the *area* of that surface and that any force that diffuses to a uniform intensity but only within the limits of a finite *three*-dimensional space will decrease inversely with the *volume* of that space. These claims are just instances of the universal law of dynamics. And this, I have suggested, might be regarded by Kant as a condition on the consistent applicability of the notion of a total consequence of a ground, as that notion is used in conceiving intensive magnitude as a quantity in the full sense – at least in certain of the most fundamental cases, namely, those that are, as Kant puts "limited . . . only by the magnitude of the space into which it must diffuse" (*MF* 4:518–519). (Kant appears to treat this as a sufficient condition for satisfying the non-derivative and the causal immediacy requirements discussed in the preceding section.) Insofar as this can be considered a general condition on applying the category of quantity to an intensive magnitude, this "universal law of dynamics" might be considered transcendental in character or, perhaps, a bit of the metaphysics of objects of outer sense (i.e., objects in space).

Second, Kant relies on the claim that if, as in the former case, these surfaces are concentric spheres, then these areas will vary inversely with the square of their radii, and, as in the latter case, where we consider the diffusion of the force to be limited within the boundaries of a three-dimensional space, then, if these spaces are spheres, then the volume varies inversely with the *cube* of their radii. These two claims are essentially geometrical in character.

Now, in neither of the two considerations just noted is there any mention of matter's actual attractive and repulsive forces. Thus, a third consideration Kant must in addition rely on is that attractive force diffuses on the two-dimensional space of a spherical surface centered at the attracting point, and repulsive force diffuses through the three-dimensional bounded volume that it fills. It is perhaps here that we can find a role for something empirical in this otherwise rather aprioristic account. It must be granted, however, that Kant's text gives us very little to work with that is at all definite with regard to where experience might fit into this account. And to some extent we must rely on our "best guess," and we have to admit that some of what follows can be advanced only in the spirit of a hypothesis advanced to save the assumption (which may be false) that Kant has a consistent story he can tell, a story that can plausibly reconcile the aprioristic aspects of the arguments we have been examining with the indispensable role for experience, which Kant repeatedly emphasizes. For certainly, in the Critical period (and probably as early as 1762–3), Kant is committed to the claim that, ultimately, the existence of forces with such and such characteristics, or even that such forces are, as Kant puts it, "really possible", is not something that can be known a priori; they can be known only with the contribution of "the data of experience."

Now, as I understand it, the *MF*'s Dynamics chapter, at least in the official sequence of its Explications and its Propositions and their Proofs, is meant to establish a priori that anything endowed with a repulsive force must also be endowed with an attractive force that acts immediately at a distance on any other body no matter how distant. The existence of (or "real possibility") of such attractive forces can be known once we grant the existence (or at least the "real possibility") of repulsive forces, but this latter, I take it, is not something that can be known a priori. Moreover, it seems to me that the inverse-square character of this attractive force is not established (or intended to be established) in the Explications and Propositions of the Dynamics chapter. And if the "data of experience" are indispensable to knowing that attractive force is governed by an inverse-square function, then it seems unsurprising that it has no place in the *MF*, within this series of a priori claims. Earlier in his development, in the 1756 *Physical Monadology*, Kant claimed that it is the *direction* of the force, toward the center or away from the center, which determines that attractive force diffuses over a two-dimensional space, whereas repulsive force diffuses over a three-dimensional space.[23] In the *MF*, even in the Notes and Remarks that (on my reading) lie outside the a priori sequence of the Dynamics' seven Explications and eight Propositions, we don't see an

[23] In the *Physical Monadology*, Kant begins his argument for the inverse-cube law with "since the repulsive force is exercised in an outward direction from the central point of the space occupied by an element . . ." (1:484). And he begins the argument for the inverse-square law with "On the other hand, since attraction is, of course, the action of the same element, albeit in the opposite direction . . ." (ibid.).

argument for the inverse-square law quite of the form seen in the pre-Critical essay.[24] Moreover, it is hard to see that any considerations Kant brings out in the passages of the Dynamics chapter, where he does discuss the grounding of the inverse square law, could be sufficient for establishing that, for attractive force, it is a *two*-dimensional (spherical) space to which the universal law of dynamics should be applied.[25] For knowledge that attraction obeys an inverse-square law, I take it, Kant thinks we need to look at the effects of this force, viz., the changes of motion that it produces (cf. A207/B252). And that will, for example, mean appealing to Kepler's laws, or extensions of Kepler's laws (such as the law that orbits can be not only ellipses but other kinds of conic sections as well). But the appeal to such laws, insofar as it is required for a *justification* of the inverse-square law, will presumably be (at least in part) an appeal to an *empirical* ground.

9.7 The Explanatory Role of the a Priori Considerations

However, this raises a question. If our knowledge that attraction obeys an inverse-square law requires appeal to observationally known regularities, like Kepler's laws, why even bother with these a priori considerations about the diffusion of forces in various spaces? Hasn't that kind of reflection become completely superfluous in light of what we know empirically from astronomers like Kepler? I think it has *not* become superfluous. I think that we can better see the role of these reflections on the diffusion of force if we focus on what Kant has to say about explanatory grounds and the way they figure in larger hierarchies of grounds. As Kant points out in the

[24] Two passages, which appear after the Dynamics chapter's official sequence of Explications and Propositions, need to be taken into consideration here. First, in discussing the inverse-square law in the *MF* (4:519, in the paragraph labeled "2"), Kant does rely on considerations very similar to those brought into the *Physical Monadology*, but only in the context of arguing about the correct and incorrect ways of representing attraction *by lines of force*. Second, he claims in a passage that comes a bit later (4:421, in the paragraph labeled "4") to have established that the force of attraction obeys an inverse-square law. However, in the first passage, as I understand it, when Kant considers the way attraction should be represented, he is *assuming* that the force of attraction diffuses into a two-dimensional (spherical) space. And we can plausibly take it that, in the second passage, when Kant suggests that he has argued for the inverse-square law, what he means is that he has inferred it from the *assumption* that attraction diffuses into a two-dimensional space. There seems to me no basis for thinking that in these passages he is inferring this assumption merely from the direction of attractive force (toward the center), as he does in the *Physical Monadology*.

[25] I note that even in the proof of *MF* Dynamics' Proposition 8, the infinite reach of attractive force is supposed to be established, Kant grants only that the degree of attraction decreases with the amount of space through which it diffuses, not that this is a two-dimensional space. Though even if he *had* said that it was two-dimensional, that would not amount to a claim that we have a priori grounds for saying so or that its two-dimensionality was essential to the proof.

Prolegomena, Section 38, where he discusses the "sources" of the inverse-square law in the properties of spherical surfaces, as well as the "consequences therefrom,"[26] the inverse-square law is the explanatory ground, and the different kinds of orbits – elliptical, parabolic, and hyperbolic – are each specific "consequences" of this more general ground. But the inverse-square law itself can in a way be explained by a higher and still more general ground.

Different force laws can be thought of as different ways in which forces can be conceived as diffusing through a space, keeping the "total quantum" of its effects unchanged. The genus is the idea of a force whose strength is inversely proportional to the space through which it diffuses. Simplifying a bit, we can say that one species is a force that diffuses through a two-dimensional space, and the other, a force that diffuses through an enclosed three-dimensional space. In this way, the inverse-square law is not simply left as a brute fact. When it is characterized as a way of diffusing through a certain kind of space, it is characterized as an instance of a more general phenomenon. And it can be seen in relation to other possible fundamental force laws, which correspond to other possible ways in which a force can be considered to diffuse in a space. It allows us to see basic force laws as derived from possible specifications of the generic idea of a force that, as Kant says, "is limited, as to the degree with which it exerts moving force on any given point at a certain distance, only by the magnitude of the space into which it must diffuse so as to act on this point" (*MF* 4:519). Given this generic characterization, we can come up with various mathematically possible specific force laws, each determined by the properties of the space into which the force is represented as diffusing. However, this does not yet tell us which of these species of force are real. For that we need, in addition, the data of experience.

What Kant calls the "universal law of dynamics" concerns forces whose action (i.e., effect) is "exerted by a point on every other point external to it . . . act[ing] immediately on this point"[27] (*MF* 4:522). It claims that *all* such forces fall under the genus described above, namely, those forces for which the "total quantum" of the immediate effects *cannot* differ at different distances, but can only be distributed in different ways over larger or smaller spaces. As a claim about "*all* such forces," this goes beyond what was stated already in the preceding paragraph. I take it that Kant regards

[26] Namely, "that all possible orbits of the celestial bodies are conic sections" (*Prol* 4:321).

[27] I mentioned earlier that such forces are to be considered fundamental. I also note that when Kant states the universal law of dynamics here, he limits the claim to *moving* forces.

the universal law of dynamics as necessary and as known a priori. However, we can still point to an indispensable role that experience must play. It is needed if we are to know which basic forces there are.[28]

9.8 The Role of Experience: Disanalogies between Attractive and Repulsive Forces

In the previous section I attempted to bring out a respect in which there is a parallelism between the inverse-square law of attraction and the inverse-cube law of repulsion. They can be seen as specifications of a single generic law. But I also want to mention a disanalogy. This lies not on the side of their ground but on the side of their consequences. For the law of attraction, the consequences that follow from it are Kepler's laws and their extensions. They can be thought of as confirming the existence of a force limited "only by the magnitude of the space into which it must diffuse," in such a way as to obey the inverse-square law. But it is not clear that, in the case of the *repulsive* force, there is the same kind of direct evidence of the activity of an inverse-cube law. One important feature of Kepler's laws is that they approximate the *immediate* effects of the corresponding force. By contrast, matter's original repulsive force is, according to Kant, a contact force. So its *immediate* effects and their diminution with distance occur only in an infinitesimally small neighborhood surrounding the repelling matter,[29] and so are not accessible to direct observation.

So, perhaps Kant thought that we could infer the one-over-*r*-cubed law of repulsion in a more *indirect* way from observable regularities. If we consider the nonimmediate effects observable in a finitely large bulk of matter, there is, for example, the phenomenon of the gas laws, specifically the law that pressure varies inversely with volume (at least if temperature is constant), which is variously called Boyle's or Mariotte's law. But, as Kant points out[30] (citing Newton's proof to this effect[31]), this would be evidence *not* of a force between component particles that varies inversely with the *cube* of the distance; rather it is evidence of a force that varies as *one-over-r*.

[28] I am not sure that Kant is really committed to the universal principle of dynamics stated in full generality, i.e., for all fundamental (moving) forces acting immediately from a given point on all other points. In the context of the *MF*, there may well be an implicit limitation to those fundamental forces that are "essential" to matter as such, those forces that *must* belong to something insofar as it fills a space to a determinate degree.

[29] Kant describes the fundamental repulsion between bits of matter as varying "in inverse cubic ratio to their *infinitely small distances*" from one another (*MF* 4:522, emphasis added).

[30] *MF* 4:522. This is also discussed by Pollok 2001, pp. 329–39.

[31] *Principia Mathematica*, Bk. II, Scholium to Prop. 23.

The problem with relying on this gas law is that the pressure of a gas that results from its compression represents what Kant calls "derivative elasticity," rather than the "original elasticity" associated with the inverse-cube law. The gas law, according to Kant, reflects an interaction of caloric fluid with the ordinary matter of air, which it permeates. And a bit later in *MF*, chapter 2 (4:530), Kant suggests that the elasticity of this caloric fluid is perhaps "original." But it should be noted that caloric fluid is typically regarded as capable of permeating any kind of ordinary matter, and so it could not be subjected to the kinds of experiments involving compression within a closed space that underwrote the gas law discovered by Boyle and Mariotte. Thus, it is a lot less clear what Kant regards as the "data of experience" relevant to grounding the inverse-cube law for repulsion than it is in the case of attraction.

The argument Kant presents in the *MF* (4:520–521, paragraph labeled "3") for the inverse-cube law of repulsion seems to depend only on the idea that the space in which this outwardly directed force is diffused is a corporeal space, which is to be thought of as filled ("*ein körperlicher Raum ist, der als erfüllt gedacht werden soll*").[32] If this is right, then, according to Kant, the fact that the fundamental repulsion under consideration is to account for the "filling" of space is sufficient to establish that it diffuses through a three-dimensional space, and thus (in accordance with Kant's "universal law of dynamics") obeys an inverse-cube law. This, however, is assumed or argued for right at the beginning of the Dynamics chapter. Once Kant has established that matter's property of "filling" space is by means of a fundamental repulsive force, he appears to have all he needs to commit himself to an inverse-cube law – even if that commitment "no longer belongs to metaphysics" (4:517) or to the proper tasks of the *MF*. If this is so, then the only role for the "data of experience," as regards the fundamental *repulsive* force, is to establish the actuality (or the "real possibility") of a force responsible for the filling of space. To be sure, Kant does build a great deal into the notion of "filling" a space, and this would be a plausible starting point for objecting to Kant's account. But in any case, the point would be that for Kant, no *further* appeal to experience, beyond that presupposed at the outset of the Dynamics chapter, is required to support the inverse-cube character of this force. And in this respect there is a disanalogy with Kant's treatment of attractive force in the *MF*. The argument of the Dynamics chapter is that if matter is endowed with repulsive force, then it must also be endowed with an attractive force.

[32] ". . . a volume [*ein körperlicher Raum*] which is supposed to be thought as filled" (4:520).

And this attractive force must act immediately at a distance, and on all bodies, no matter how distant. The reality of the repulsive force can be given only through experience. But once that is given, the reality of this attractive force is guaranteed as well. What I am suggesting here is that, in the case of attractive force, unlike repulsive force, there is a *further* role for "data of experience" in establishing the specific mathematical form (the inverse-square character) of the relevant force law.

Acknowledgments

I wish to thank the editors of this volume for their patience and their helpful comments and suggestions. I am also grateful to audiences who heard earlier versions of this chapter at conferences at the Hebrew University in Jerusalem and at a workshop at Stanford University. In particular, I would like to express my gratitude to Michael Friedman and Eric Watkins for helpful questions and discussion. Most of all I am thankful to Hannah Ginsborg for her advice and encouragement.

PART IV

Laws in Physics

CHAPTER 10

Kant's Conception of Causal Necessity and Its Legacy

Michael Friedman

My previous discussions of Kantian causal necessity have emphasized the role of the category of necessity in the Postulates of Empirical Thought. I have argued, against the views of Gerd Buchdahl and others, that Kant does have an anti-Humean conception of necessary causal connections and necessary empirical causal laws and that this kind of necessity involves an essential *constitutive* contribution from the faculty of understanding: merely *regulative* contributions from the faculties of reason and reflective judgment are insufficient. I have also argued that the further articulation of Kant's conception from the first edition of the *Critique of Pure Reason* (1781) through the *Prolegomena* (1783) and *Metaphysical Foundations of Natural Science* (1786) to the second edition (1787) shows that the Newtonian law of universal gravitation is paradigmatic, for Kant, of a necessary but still empirical causal law.[1]

I am now extending my interpretation by focusing on the category of causality and, in particular, on the way in which this category is related to the "predicable" (derivative category) of force. I argue that Kant's conception of force is more indebted to the Newtonian mathematical concept of impressed force than I had previously emphasized and that what Kant takes to be necessary and strictly universal empirical causal laws – such as, paradigmatically, the law of gravitation – thereby acquire a more than inductive and more than hypothetical epistemic status. Newton's methodology of "deduction from phenomena" thus turns out to be central to Kant's conception of causal necessity. I also then explore the legacy of this conception – the prospects for extending it into the future in the context of post-Newtonian physical theories such as Einstein's theory of relativity.

[1] My first formulation of this interpretation was Friedman 1992b. My most developed recent discussion is Friedman 2012b. See also Friedman 2014 for a summary of this last discussion, in connection with the second edition Transcendental Deduction and the third *Critique*. Buchdahl 1969, by contrast, argues that the Transcendental Analytic is not concerned with the necessity of empirical causal laws, which depends solely on reason's regulative demand for systematicity.

10.1 Force in Kant and Newton

I develop my interpretation around a new reading of the Preface to the *Metaphysical Foundations of Natural Science*.[2] I begin with the fourth paragraph of the Preface, where, after introducing the notion of *proper* natural science in the third paragraph, Kant explains that this kind of science depends on its pure part (*MF* 4:468–469):

> Since the word nature already carries with it the concept of laws, and the latter carries with it the concept of the *necessity* of all determinations of a thing belonging to its existence, one easily sees why natural science must derive the legitimacy of this title only from its pure part – namely, that which contains the a priori principles of all other natural explanations – and why only in virtue of this pure part is natural science to be proper science.

Kant thus indicates that "proper" natural science – paradigmatically physics – is wider than "pure" natural science: the latter is strictly a priori, and it makes possible empirical natural laws, like the law of gravitation, which are necessary (in the sense of the category of necessity) but not strictly a priori. Kant then contrasts chemistry (implicitly) with physics. Explanation on the basis of chemical principles "still always leaves behind a certain dissatisfaction, because one can adduce no a priori grounds for such principles, which, as contingent laws, have been learned merely from experience" (469). In the case of physics, however, the a priori grounds in question belong to *pure* natural science, but there are also empirical natural laws, like gravitation, which still count as necessary and belong to *proper* natural science.[3]

All demonstrated Propositions in the *Metaphysical Foundations* – especially Kant's three Laws of Mechanics (conservation of quantity of matter, inertia, action equals reaction) – are strictly synthetic a priori. But the law of universal gravitation is not demonstrated a priori, for Kant, but is inferred from Kepler's so far merely inductive "rules" by what Newton calls "deduction from phenomena."[4] The law of universal gravitation – and thereby Kepler's laws as well (as suitably *corrected* by Newton) – now

[2] I discuss this Preface at length in Friedman 2013. The present discussion, however, is almost entirely new.

[3] Kant has already indicated that chemistry is not a "proper" natural science at the end of the third paragraph (468). The contrast with physics in general and the Newtonian law of gravitation in particular is made explicit in the eighth paragraph, discussed below.

[4] Kant's clearest statement in the *Metaphysical Foundations* occurs in the General Remark to Dynamics (534): "[N]o law of either attractive or repulsive force may be risked on a priori conjectures. Rather, everything, even universal attraction as the cause of weight [*Schwere*], must be inferred, together with its laws, from data of experience." Kant considers the transition from Kepler's "rules" to Newtonian "laws" in a number of texts, some of which are addressed in Friedman 2012b: 313–314.

acquire a more than merely inductive status. I elaborate on this status in what follows.

Consider a passage from the second remark to Proposition 7 of the Dynamics where Kant appears sharply to differ with Newton concerning gravity as an immediate action at a distance (515): "[I]t is clear that the offense taken by his contemporaries, and perhaps even by Newton himself, at the concept of an original attraction set him at variance with himself: for he could absolutely not say that the attractive forces of two planets, for example, of Jupiter and Saturn, manifested at equal distances of their satellites (whose mass is unknown), are proportional to the quantity of matter of these heavenly bodies, if he did not assume that they attracted other matter merely as matter, and thus according to a universal property of matter." At issue is the method Newton develops in Book 3 of the *Principia* for measuring the masses of primary bodies by the accelerations produced in their satellites: Newton, according to Kant, must assume the immediacy of the gravitational interaction between Jupiter and Saturn, independently of any intervening matter surrounding them, in order to infer their masses from the accelerations of their moons.[5]

Is Kant saying that Newton must accept the existence of real action at a distance despite the qualms vigorously expressed by contemporary representatives of the mechanical philosophy such as Huygens and Leibniz? This question is more subtle than I had previously thought. For consider what Kant says before the just-quoted passage (515, emphasis added): "[Newton] *rightly* abstracted from all hypotheses purporting to answer the question as to the cause of the universal attraction of matter, for this question is physical or metaphysical, but not mathematical."[6] Here Kant is endorsing the abstract treatment of impressed force as a measurable quantity that Newton has been articulating since Proposition 1 of Book 1, which demonstrates that Kepler's area law for a given trajectory is mathematically equivalent to generation by a centripetal force (always directed toward a single center), whatever its deeper explanation might be.[7] The aim is to determine the mathematical properties of this force,

[5] I first developed this point in Friedman 1992a, and I recur to it in most of the works cited above; see also Friedman 2012c.

[6] In using the disjunction "physical or metaphysical" Kant echoes Newton's *hypotheses non fingo* passage quoted below. It is important to note, however, that Kant differs from Newton regarding the need for metaphysics in the foundations of physics; see Section 10.3 below.

[7] The treatment is thus "abstract" in the sense of *abstracting from* any such explanation; it does not mean that the force is an abstract mathematical entity in the sense of contemporary mathematical Platonism (more on this below).

and then to use these properties to support inferences from empirical (albeit mathematically expressed) propositions like Kepler's "rules" to further mathematical properties of the force in question (such as the inverse-square law for gravitational force). We thereby open the way to a continued series of empirical inferences that progressively correct the mathematical laws obtained at one stage by then considering ever-more subtle effects of gravitational force, which both further support theseearlier-obtained laws while also revealing their fundamentally approximate character.

Thus, for example, Newton's derivation in Proposition 3 of Book 3 of an inverse-square acceleration of the moon toward the Earth involves a promissory note concerning the perturbative effect of the Sun on the Moon. But the note begins to be redeemed, later in this Book, only with Newton's detailed attack on the three-body problem, and it is fully redeemed, after Newton's death, only by Euler's and Clairault's first successful treatment of the motion of the Moon in the 1740s. A similar sequence of increasingly accurate approximations leads (as already suggested) from Kepler's "rules" assumed as exact to corrections of them based on the perturbative effects of bodies other than the Sun, and so on. This procedure of approximative correction and theory-mediated measurement continues with ever-increasing success throughout the nineteenth century, the most spectacular being the well-known discovery of Neptune by Adams and Leverrier in 1846.[8] Decisive obstacles arose only at the turn of the twentieth century, when minute deviations in the predicted advance of the perihelion of Mercury eluded all attempts at solution within Newtonian theory. The first successful solution then became one of the triumphs of Einstein's general theory of relativity, a fundamentally revised theory of gravitation, in 1915. I return to this case below, in discussing the legacy of Kant's conception in our post-Newtonian context.

But what is the evidence for attributing the Newtonian conception of force – which abstracts from all hypotheses concerning the underlying mechanism of gravitational interaction – to Kant? To begin with, Kant was undoubtedly familiar with Newton's *hypotheses non fingo* passage in the General Scholium to the *Principia*: "I have not as yet been able to deduce

[8] The idea of a characteristically "Newtonian style" of combining empirical and mathematical reasoning in a progression of successive approximations is due to I. B. Cohen (1980). It has more recently been elaborated and extended by George E. Smith (2002, 2012, 2014), who has given particular emphasis to the notion of "theory-mediated measurement" – where fundamental principles of the theory (here the Newtonian Laws of Motion) enable robust empirical measurements of its fundamental physical quantities (such as mass, force, and acceleration).

from phenomena the reason for these properties of gravity, and I do not feign hypotheses. For whatever is not deduced from the phenomena must be called a hypothesis; and hypotheses, whether metaphysical or physical, or based on occult qualities, or mechanical, have no place in experimental philosophy."[9] Newton is here defending his mathematical law of gravitation from criticisms raised by mechanical philosophers (especially Huygens and Leibniz), who insisted on the priority of action by contact in order to explain this law, and Kant develops a parallel response to the mechanical philosophy on behalf of what he calls a "metaphysical-dynamical mode of explanation [*Erklärungsart*]" in the General Remark to Dynamics.[10]

The most significant evidence, however, is found in the discussion following the fourth paragraph of the Preface (468–469). In the fifth paragraph Kant explains that the *pure* part of proper natural science consists of two further parts (469): "pure rational cognition from mere *concepts*" or "metaphysics," and "that which grounds its cognition only on the *construction* of concepts" or "mathematics." He asserts that we must presuppose, in the first place, "metaphysics of nature" – "[f]or laws, that is principles of the necessity of that which belongs to the *existence* of a thing, are concerned with a concept that cannot be constructed, since existence cannot be presented a priori in any intuition" (ibid.). But "metaphysics of nature," according to the sixth paragraph (469–470), again consists of two further parts: a metaphysics of nature *in general*, which considers only the pure concepts of the understanding, and a *special* metaphysics of nature, which, for example, considers the empirical concept of matter as a particular instantiation of the pure concepts of the understanding.

There then follows a famous, and very difficult (seventh) paragraph, according to which special, unlike general, metaphysics *requires* the application of mathematics. The crux of the difficulty involves the role of mathematical construction:

9 Quoted from Cohen and Whitman 1999: 943.

10 The passage quoted in note 4 above concerning the law of gravitation is an important part of this discussion. Kant's contrast between the "mathematical-mechanical" and "metaphysical-dynamical" modes of explanation focuses on the need to assume empty interstices to account for variations in specific densities (a theory of atoms and the void). Yet this discussion is intertwined with a second theme concerning whether "fundamental forces" can be introduced hypothetically without certification of their (real) possibility in experience. In criticizing the hypothetical method of the mathematical-mechanical approach, Kant says (532) that "here is not the place to uncover hypotheses for particular phenomena, but only the principles in accordance with which they all are to be judged. Everything that relieves us of the need to resort to empty spaces is a real gain for natural science, for they give the imagination far too much freedom to make up by fabrication for the lack of any inner knowledge of nature." Kant's "fabricate [*erdichten*]" echoes Newton's "feign [*fingère*]."

> [T]o cognize something a priori means to cognize it from its mere possibility. But the possibility of determinate natural things cannot be cognized from their mere concepts; for from these the possibility of the thought (that it does not contradict itself) can certainly be cognized, but not the possibility of the object, as a natural thing that can be given outside the thought (as existing). Hence, in order to cognize the possibility of determinate natural things, and thus to cognize them a priori, it is still required that the *intuition* corresponding to the concept be given a priori, that is, that the concept be constructed. (470)

Is Kant saying that special metaphysics – in particular the special metaphysics of corporeal nature – proceeds by mathematically constructing its concept of matter in pure intuition? This cannot be right. For the concept of matter is an *empirical* concept, and Kant is clear – for example, in the Postulates of Empirical Thought – that empirical concepts as such cannot be constructed in pure intuition.[11] More telling in the present context, however, is the very passage from the Preface we are considering. The possibility that needs to be demonstrated a priori is that "of the object [of the concept], as a natural thing that can be given outside the concept (*as existing*)" (470, emphasis added). And Kant has initiated this discussion by explaining the need to begin with metaphysics of nature – as opposed to mathematics – because, in this context, we "are concerned with a concept that cannot be constructed, since *existence* cannot be presented a priori in any intuition" (469, emphasis added). Moreover, according to Kant's discussion of the Analogies of Experience in the first *Critique*, the (general) metaphysical principles governing the *existence* of things are precisely these Analogies.[12]

Kant is not saying that the empirical concept of matter can be constructed in pure intuition, independently of both general metaphysics and experience. The point, rather, is that, assuming general metaphysics as our background, we must articulate the empirical concept of matter as an instantiation of the categories of the understanding, and this must be a *mathematical* instantiation, in particular, of the categories of substance, causality, and community. Thus, the most important principles of the special metaphysics of corporeal nature are Kant's three Laws of

[11] See the contrast between pure concepts (both mathematical and categorical) and empirical concepts in the discussion of the first Postulate (A220–222/B267–270); compare the treatment of this issue in Friedman 2013: 28–29.

[12] The relevant paragraph begins (A215/B262): "[The three Analogies of Experience] are nothing else but the principles for the determination of the existence of appearances in time with respect to all of its three modes, the relation to time itself as a magnitude (the magnitude of existence, i.e., duration), the relation in time as a series (successively), and finally [the relation] in time as a totality of all existence (simultaneously)."

Mechanics, which instantiate the corresponding categories by applying pure mathematics to the empirical concept of matter. We thereby obtain a mathematically precise notion of quantity of matter (and thus quantity of substance), a precise notion of force or causal action (governed by the law of inertia), and a precise notion of mutual action or interaction (governed by action equals reaction).

It is no wonder, then, that Kant proposes precisely these Laws of Mechanics in the Introduction to the second (1787) edition of the *Critique* as indisputable examples of synthetic a priori propositions of "pure natural science."[13] It is also no wonder that, when discussing the a priori concepts of substance, causality, action, and force in the Second Analogy, Kant appeals to the example of *moving forces* to provide the particular content of an alteration of state (A207/B252): "For this acquaintance with actual forces is required, which can only be given empirically, e.g., the moving forces, or, what is the same, certain successive appearances (as motions), which indicate such forces." And, since these forces give rise to law-governed motions (accelerations) conceived as alterations of state, the law of inertia is also necessarily involved.[14]

I round off this discussion of textual evidence with the eighth paragraph of the Preface. Here Kant explains why chemistry is not yet a proper natural science, but rather a mere "systematic art or experimental doctrine" (471). This, in particular, is because for chemical actions there is "no concept to be discovered that can be constructed, that is, no law of the approach or withdrawal of the parts of matter can be specified according to which, perhaps in proportion to their density or the like, their motions and all the consequences thereof can be made intuitive and presented a priori in space" (470–471). The problem, therefore, is that there is (as yet) no attractive or repulsive mathematical force law analogous to the law of gravitation. As we have seen, the latter law is empirical rather than a priori, but it is still mathematically expressible in abstraction

[13] Kant has just asserted that one can appropriately ask *how* pure mathematics and pure natural science are possible, since "*that* they must be possible is demonstrated by their actuality" (B20), and he appends the following footnote (B20–21n.): "Some may still have doubts concerning the latter. But one need merely consider the various propositions that come forth at the outset of proper (empirical) physics, such as those of the permanence of the same quantity of matter, of inertia, of the equality of action and reaction, etc., and one will be quickly convinced that they constitute a *physica pura* (or *rationalis*), which well deserves to be separately established, as a science of its own, in its whole extent, whether narrow or wide."

[14] See the footnote appended to the main paragraph (A207/B252n.): "One should well note that I do not speak of the alteration of certain relations in general, but rather of alteration of state. Therefore, if a body moves uniformly it does not alter its state (of motion) at all, but it certainly does if its motion increases or decreases."

from all hypotheses concerning underlying explanations. This points back to the discussion of pure and proper natural science in the fourth paragraph (468–469), which concludes with the claim that chemistry (unlike physics) contains no properly necessary (albeit still empirical) natural laws.[15]

Before turning to the legacy of Kant's conception of force and causal necessity in post-Newtonian physical science, I shall summarize and clarify the relationship between this conception and Newton's. The first point is that Kant accepts Newton's abstract conception of force as a measurable mathematical magnitude, considered independently of all merely hypothetical elements. This does not mean that force is either what we would call an abstract mathematical entity or what we would call an observable as opposed to theoretical entity – one reducible to purely observational, theory-neutral concepts. On the contrary, the fundamental quantities of mass, force, and acceleration function as primitive terms in Newton's physical theory, where they are related to one another mathematically by the Laws of Motion. The latter serve to enable *theory-mediated* measurements of these quantities, including the quantity of duration (elapsed time). Thus, for example, given the observed Keplerian orbits, Proposition 1 of Book 1 provides an empirical measure of duration in terms of geometrical areas; later propositions of the *Principia* provide an empirical inverse-square estimation of the centripetal force in question; and, as we have seen, still later propositions successively build on these results to provide a universal common measure of mass or quantity of matter for the primary bodies in the solar system.[16] So this conception emphasizes empirically determined relations between empirically measurable (albeit still theoretical) quantities, rather than hypothetical postulation of theoretical entities that may not have, at least at present, any such means of empirical determination. Thus, for example, while contemporary representatives of the mechanical philosophy insisted on the need for vortex models of planetary motion to explain Newton's results, he was well within his

[15] Compare note 3 above, together with the paragraph to which it is appended. The eighth paragraph concludes (471): "[C]hemical principles do not in the least make the principles of chemical appearances conceivable with respect to their possibility, for they are not receptive to the application of mathematics." This confirms the view that application of mathematics to empirical laws in the Newtonian style, rather than a priori construction of natural scientific concepts and principles, is central to Kant's conception of proper natural science.

[16] See the paragraph to which note 6 above is appended. Kepler's area law, discussed in the following paragraph, says that equal areas of a Keplerian orbit are swept out in equal times. An account of Newton's more general derivation of this law – valid for any trajectory generated by a centripetal force – can be found in Friedman 2013: 371–372.

rights to reject this insistence on behalf of his mathematical law of gravitation – since the vortical motions in question lacked all means, at the time, of robust empirical determination.[17]

The possibility of robust empirical determinations of the relevant physical quantities provides Newton's theory with a more secure epistemic warrant than either mere inductive inferences (curve-fitting) or mere hypothetico-deductive arguments (inference to the best explanation). The relative weakness of mere hypothetico-deductive arguments is well illustrated by the example of Newton's confrontation with the mechanical philosophy over its insistence on vortex models of planetary motion. For the relative weakness of mere inductive inferences, however, consider the relationship between Newton's argument for the law of gravitation and Kepler's (so far) merely inductive "rules." To be sure, Newton begins his argument from the observable Keplerian phenomena, taken provisionally as well-supported universal claims. The crucial point, however, is that Newton's argument involves an open-ended sequence of *corrections* to Kepler's "rules" taking increasingly accurate account of gravitational perturbations. Indeed, these corrections lead to the conclusion that the "orbits" in question are not even closed curves (due to perturbatively induced precession) – a conclusion that could not be effectively reached by the collection of further data points and curve-fitting. It is precisely here, in my view, that the greater strength of the characteristically Newtonian method of reasoning from phenomena becomes fully clear.[18]

10.2 The Legacy of Kant's Conception in Post-Newtonian Physics

The original Newtonian method breaks down, as suggested, in the case of minute deviations from the predicted perturbations of the orbit of Mercury, which are satisfactorily explained only on the basis of Einstein's

[17] This, in fact, was one of the main points at issue in Newton's *hypotheses non fingo* passage. For Leibniz's attempt to provide a vortex model for Newton's treatment of centripetal force (and Newton's reaction), see Bertoloni Meli 1993.

[18] Compare note 10 above, together with the paragraph to which it is appended. An important advantage of this method is that Newton could settle the choice between geocentric and heliocentric world systems empirically by theory-mediated measurements of the masses of the primary bodies in the solar system: the center of mass of the system (which is always very close to the center of the Sun) is thereby determined as the preferred dynamical center for describing the true motions. A merely kinematical description, by contrast, involves an arbitrary choice of reference frame. See the works cited in notes 1 and 2 above for further discussion. (It would be more accurate not to describe Kepler's reasoning as mere inductive curve-fitting, but this oversimplification nonetheless serves to bring out the strength of Newton's method.)

theory of relativity developed in the years 1905–1915. At this point our inquiry takes a completely unexpected direction by now considering a non-Newtonian causal action exerted by the variable non-Euclidean curvature of space-time. And it is at precisely this point that the original Kantian conception of force and causal necessity also breaks down. Kant does have room for the corrections to the planetary orbits that can be accounted for within the Newtonian theory of gravitation – that is, the Newtonian gravitational perturbations.[19] But he has no room for causal action exerted by a non-Euclidean variation of curvature in our form of outer (spatial) intuition. So the question of the legacy of Kant's conception becomes that of the precise way in which it might now be appropriately extended and generalized.

I shall approach this question, as before, by focusing on the concept of force. Einstein's general theory of relativity is, like Newton's theory, a theory of specifically gravitational force. In order properly to appreciate this theory, however, we must first gain an appreciation of Einstein's earlier (1905) special theory of relativity, which is a post-Newtonian spatiotemporal framework for considering electromagnetic force – the second of what we now take to be the four fundamental forces of nature, beginning (in order of discovery) with gravitational force.

An adequate mathematical description of what we now call the electromagnetic field was achieved by Maxwell in the second half of the nineteenth century, in terms (as we now understand it) of electric and magnetic force vectors continuously distributed over space. This field of force, like Newtonian gravitational force, is a mathematically described physical quantity. And, as such, it can be studied in abstraction from (independently of) any deeper physical realizations – involving, for example, underlying motions in mechanical, fluid, or elastic ethers. Unlike Newtonian gravitational force, however, the electromagnetic field is *dynamical*: it is propagated through space with a definite finite velocity c (the velocity of light). It is thus, as Howard Stein (1970) has emphasized, an actor in the physical world along with the matter on which it acts: there are energy and momentum exchanges between matter and the field, for example, not only among the material bodies themselves.

[19] Kant acknowledges such corrections in the Appendix to the Dialectic (A663/B691): "[W]e arrive at unity of the genera of these paths according to their form [at circular, elliptic, parabolic, and eventually hyperbolic orbits]; and we thereby further arrive, however, at unity of the cause of all the laws of their motion (i.e., gravitation). From there we afterwards extend our conquests further, seeking also to explain all variations and apparent deviations from these rules from the same principle."

Light waves, on Maxwell's theory, are propagations in the electromagnetic field confined to wavelengths within the visible spectrum. But the electromagnetic spectrum extends beyond visible light on both ends: with successively shorter wavelengths, as in ultraviolet, X-rays, and gamma rays, and successively longer wavelengths, as in infrared, microwave, and radio waves. Hertz's celebrated experiments toward the end of the nineteenth century involved the first generation and reception of electromagnetic waves by means of what we would now call a radio transmitter and receiver located several meters apart. Hertz could thereby measure the velocity of transmission (which turned out, as predicted, to be the same as the velocity of light) and compare the frequencies of oscillations in the transmitter and antenna receiver (which turned out, as predicted, to be the same). These techniques eventually enabled the experimental manipulation of such "electric waves" by reflections, refractions, and so on, just as we had long done with light waves. Hertz's experiments, in this sense, empirically realized the electromagnetic field and thereby established its mathematical properties, just as those of gravitational force had been established two centuries earlier by Newton's inferences from phenomena.[20]

In both cases, however, "established" does not mean proved to be correct with no possibility of revision. Rather, Newton's inferences from phenomena and Hertz's experiments provided the corresponding forces with observationally determinable empirical reality. They thereby established the possibility of continuing progressive empirical inquiry into these forces – a process that, as we saw, is always open to further correction as our empirical determinations of the relevant quantities become increasingly precise. This kind of reasoning provides such empirical determinations with greater epistemic security than either merely inductive reasoning or hypothetico-deductive reasoning. Thus, just as in the case of Newton and the mechanical philosophy, the hypothetical ether models that were originally proposed for both the wave theory of light and Maxwell's theory were considerably more problematic than Hertz's experimental investigations into the mathematical properties of the electromagnetic field – which could thus be empirically and mathematically investigated independently of any such model.

[20] Hertz's experimental work is the basis for all wireless technology, beginning with the invention of the radio by Marconi and extending to contemporary cellular (microwave) technology. For recent historical accounts of Hertz's work – showing, among other things, that its relationship to Maxwell's theory is considerably more complicated than suggested above – see Buchwald 1994; Darrigol 2000.

We can now begin to see how Einstein's general theory of relativity is bound up with his special theory by observing that there is a fundamental tension between the kind of causality exemplified by the electromagnetic field and that of Newtonian universal gravitation. The latter is propagated instantaneously, along "planes of absolute simultaneity" in the Newtonian spatiotemporal background structure, while the latter is always propagated with finite velocity *c*. But no finite velocity can be the same in two different inertial frames of reference in a Newtonian spatiotemporal structure – where an inertial frame is one in which the law of inertia is valid (together with the other Newtonian Laws of Motion). Only the planes of absolute simultaneity themselves, representing infinitely fast or instantaneous causal actions, are invariant in such a structure. The special theory of relativity then addressed this tension by changing the spatiotemporal background to what we now call "Minkowski space-time" – wherein the mathematical properties of Maxwellian (finitely propagated) causality are taken to be those of the spatiotemporal background structure as well. In particular, the simultaneity relation is *relativized* so that there are different planes of simultaneity through a given point-event relative to different inertial frames. The "light cone" determined by all spatiotemporal trajectories with the given finite velocity through a given point-event now replaces the plane of absolute simultaneity (through this point-event) as fundamental spatiotemporal invariant.[21]

Interferometer experiments by Michelson, Morley, and others at the end of the nineteenth and beginning of the twentieth century securely established these mathematical properties of the spatiotemporal background relative to those of the electromagnetic field – that is, the properties of what we now call Minkowski space-time – in the same sense that the mathematical properties of the electromagnetic field were previously established by Hertz. By contrast, appealing instead to hypothetical physical mechanisms generating contractions of rods and retardations of clocks while retaining the Newtonian spatiotemporal background goes well beyond what can be empirically established (at least so far) in the same way that Newton's opponents within the mechanical philosophy appealed to (otherwise empirically undeterminable) vortices in the ether to explain the motions of the planets.[22]

[21] Intuitive diagrams depicting the crucial differences between Newtonian and Minkowski space-time can be found in Friedman 1983. See also the beautifully intuitive treatment in Geroch 1978, viewable on Google Books.

[22] A clear and fairly detailed discussion of the Michelson–Morley (and related) experiments can be found in Resnick 1968. For a more historically oriented discussion, including of the Lorentz–Fitzgerald contraction hypothesis and other such "compensatory" hypotheses, see Darrigol 2000.

The crucial question for Einstein's general theory of relativity was then how to represent the gravitational field within the background structure of Minkowski space-time – where this field must now be finitely propagated like the electromagnetic field. And the solution Einstein arrived at was to retain the structure of Minkowski space-time – and thus the light-cone structure – as far as possible, while simultaneously introducing a *variable curvature* into the underlying space-time manifold on which the light-cones are defined: in particular, the orientations and dimensions of the light-cones can vary from point to point depending on the space-time curvature. This variable light-cone structure then determines the two possible types of (four-dimensional) trajectories representing geodesic motions in a gravitational field – that is, *straightest possible* curves within the geometry in question. In particular, the time-like geodesics always lie within the light-cones and represent freely falling (massive) bodies subject to none but gravitational interactions, while the light-like geodesics lie on the surfaces of the light-cones themselves.[23]

The resulting four-dimensional space-time geometry has implications for three-dimensional, purely spatial geometry as well. Indeed, one can visualize the purely spatial geometry of the solar system in two dimensions by imagining that the Sun is a massive ball placed on a rubber sheet. The resulting geometry is circularly symmetric with increasing negative curvature as one radially approaches the central Sun, and this representation allows one to visualize Einstein's prediction of the deflection of light in the gravitational field of the Sun that was tested during eclipse observations in 1919.[24] To fully appreciate the import of these observations, however, one needs to appreciate that the Einsteinian prediction was being tested against a corresponding Newtonian prediction (going back to the turn of the nineteenth century), whereby light was represented as a very small (and very light) massive particle moving at the (already known) velocity *c*. In particular, the Newtonian predicted deflection turned out to be one-half the value of Einstein's. Thus, the observations and photographic measurements of the full Einsteinian deflection in 1919 empirically established the mathematical relationship between Newtonian and Einsteinian gravity (where gravity is represented as a curvature imparted to Minkowski space-time rather than as a Newtonian impressed force), just as the

[23] For further discussion of the resulting general-relativistic space-time structure, including intuitive diagrams, see again Friedman 1983; Geroch 1978. The light-cone diagram on the cover of the latter depicts the space-time structure within and around a black hole.

[24] See the diagram in Misner, Thorne, and Wheeler (1973: 32), also viewable on Google Books.

previous observations and experiments of Newton, Hertz, and Michelson and Morley et al. established the mathematical properties of the different forms of causal interaction at issue in these cases.[25]

The continuity in method and resulting epistemic status involved in all three cases captures the sense in which Kant's conception of causal necessity can indeed be extended into the post-Newtonian period. Of course the nature of the causal relation undergoes important changes. In the Newtonian case the relevant force – the gravitational field – mediates the interactions between material substances without being an actor in turn. The only substances, properly speaking, are massive material bodies, and they do not exchange momentum, in turn, with the gravitational field. This situation fundamentally changes with the electromagnetic field, which is continuously propagated with a finite velocity rather than instantaneously; it thereby exchanges both momentum and energy with the relevant (charged) bodies on which it acts and is thereby acted upon. The situation changes even more radically in the general theory of relativity, where the gravitational field is not only propagated continuously but is now a field of geometrical curvature in the fabric of space-time itself. And this transformation of geometry – including even the three-dimensional geometry of space – into a physically acting and interacting field of curvature implies that Kant's conception of space as the a priori form of outer intuition has most definitely broken down. Nevertheless, a central element in his conception of causal necessity – its more than inductive and more than merely hypothetical epistemic status – is in fact preserved, relatively straightforwardly, even here.

The fate of Kant's conception of pure intuition is less straightforward, but there is a natural way to *generalize* his conception of space as an a priori form of intuition so as to embrace all the post-Newtonian developments we have considered. Instead of viewing this form as reaching globally into the cosmos – so as thereby to enable Newton's Euclidean treatment of the motions of the heavenly bodies within the solar system (and, for Kant, throughout the rest of the cosmos as well) – we view it as reaching only locally into a very small region (measured in cubic meters) embracing the laboratory and/or observatory spaces of the relevant experiments. We see this, in fact, in all of our post-Newtonian cases: Hertz's experiments with radio waves, the Michelson, Morley, et al. interferometer experiments, and

[25] See Kennefick 2009 for an excellent discussion of the 1919 observations. The advance of the perihelion of Mercury and other classical tests of general relativity can be treated similarly, albeit not so simply and intuitively. For details, see Misner, Thorne, and Wheeler 1973.

the photographic observations and measurements verifying Einstein's prediction for the deflection of light. In the last case, in particular, the space of the observatory is so small relative to the corresponding space of the cosmos (here extending well beyond the solar system) that it is, for all intents and purposes, *infinitesimally* small; so its geometry, even according to the general theory of relativity, remains Euclidean.[26] In this sense, it is still reasonable to view the use of Euclidean geometry as an a priori constitutive presupposition of the empirical observations that serve as tests for the theory, even though the global (cosmic) space employed by the theory is measurably non-Euclidean.

10.3 The Legacy of Kant's Metaphysics of Causality

I have argued that Kant's conception of causal necessity centrally involves the Newtonian method of deduction from phenomena – which, in particular, depends on a mathematical description of force that abstracts from hypotheses not yet determinable from phenomena. I have also argued that this method can be seen as continuing into the post-Newtonian period in the work of Maxwell, Hertz, and Einstein. By emphasizing the mathematical and methodological aspects of Kant's conception, however, I have so far said relatively little about its metaphysical aspects. Yet it is clear from my reading of the Preface to the *Metaphysical Foundations* in Section 10.1 that Kant's "metaphysics of experience" – based on the pure categories of the understanding and pure forms of sensibility – plays a central role in his conception. In particular, the categories of substance, causality, and community, together with the three corresponding Analogies of Experience, are fundamental. This aspect of Kant's conception represents an evident divergence from Newton's more guarded attitude toward metaphysics, and so it is necessary to consider it further here. Otherwise, it might appear that what I have said about the legacy of Kant's conception concerns the legacy of Newton rather than Kant.

In my reading of the Preface in Section 10.1 I left off with the seventh and eighth paragraphs, where Kant explains the importance of the application of mathematics in special metaphysics and contrasts chemistry unfavorably with physics in terms of its status as a proper natural science.

[26] E.g., the 1919 observatory at Sobral in Brazil was less than 10 cubic meters in volume; the stars observed belong to the Hyades cluster, the center of which is approximately 150 light-years from Earth. In general, in any Riemannian manifold of any curvature and dimension, geometry in the infinitesimally small (of the tangent space at any point) is nonetheless flat or Euclidean.

The following tenth paragraph then explains how, exactly, special metaphysics is to provide an explanation for "the application of mathematics to the doctrine of body" (472).[27] For this purpose, Kant says (ibid.) that "a complete analysis of the concept of a matter in general will have to be taken as the basis, and this is a task for pure philosophy – which makes use of no particular experiences, but only that which it finds in the isolated (although intrinsically empirical) concept itself, in relation to the pure intuitions of space and time, and in accordance with laws that already essentially attach to the concept of nature in general, and is therefore a genuine *metaphysics of corporeal nature*." Thus, once again, the task of the *Metaphysical Foundations* is not to construct the empirical concept of matter in pure intuition but to embed mathematical constructions applied to empirical (physical) concepts within the (general) metaphysics of the first *Critique*.[28]

The following eleventh paragraph is of particular interest, because here Kant emphasizes a significant difference between his approach and Newton's. Kant begins by describing what he takes to be the typical attitude of mathematically inclined natural philosophers (472): "Hence all natural philosophers who have wished to proceed mathematically in their occupation have always, and must have always, made use of metaphysical principles (albeit unconsciously), even if they themselves solemnly guarded against all claims of metaphysics upon their science." Yet these natural philosophers have understood metaphysics in the wrong way (ibid.): "Undoubtedly they have understood by the latter the folly of contriving possibilities at will and playing with concepts, which can perhaps not be presented in intuition at all, and have no other certification of their objective reality than that they merely do not contradict themselves." Comparing this sentence with the argument of the difficult seventh paragraph (470) may again tempt us to think that special metaphysics, in particular, has the task of constructing its concepts in pure intuition.

That this is not the case, however, is confirmed by what Kant says next:

> All true metaphysics is drawn from the essence of the faculty of thinking itself, and it is in no way fabricated [*erdichtet*] on account of not being

[27] The intervening ninth paragraph argues that "the empirical doctrine of the soul must remain even further from the rank of a properly so-called natural science than chemistry" (471).

[28] The following paragraph of the Preface (considered below) makes it clear that the concepts (in the plural) to which mathematical constructions are to be applied are those "of motion, the filling of space, inertia, and so on" (472) – i.e., the (constituent) concepts yielded by "the complete analysis of the concept of a matter in general" (ibid.) mentioned in the tenth paragraph. Compare the discussion of these same concepts – in relation to both the table of categories and the four chapters of the *Metaphysical Foundations* – in the fifteenth paragraph (476–477).

> borrowed from experience. Rather, it contains the pure actions of thought, and thus a priori concepts and principles, which first bring the manifold of *empirical representations* into the law-governed connection through which it can become *empirical* **cognition**, that is, experience. Thus these mathematical physicists could in no way avoid metaphysical principles, and, among them, also not those that make the concept of their proper object, namely matter, a priori suitable for application to outer experience ... But they rightly held that to let merely empirical principles govern these concepts would in no way be appropriate to the apodictic certainty they wished their laws of nature to possess, so they preferred to postulate such [principles], without investigating them with regard to their a priori sources. (472)[29]

Thus "true metaphysics" again refers to Kantian metaphysics, which is concerned with "*empirical* cognition, that is, experience" (emphasis altered). The relevant "mathematical physicists" are dissatisfied with the merely hypothetical "fabrications" characteristic of the mechanical philosophy, and Kant's investigation of the "a priori sources" of their fundamental principles in the *Metaphysical Foundations* is intended to fill in the lacuna in their purely postulational approach. There can be very little doubt, therefore, that Newton is paradigmatic of the physicists in question, and, for Kant, the task left open for his metaphysical treatment is, among other things, that of explaining the a priori character of the "Axioms or Laws of Motion" initiating the *Principia*.[30]

The required metaphysical explanation, as we have seen, proceeds by showing how the fundamental empirical concepts of Newtonian physics – such as mass, force, motion (acceleration), and interaction – are mathematically precise instantiations of the categories of the understanding, in particular, the categories of substance, causality, and community.[31] The categories of the understanding, in turn, derive from the logical structure of the pure intellect, and, when applied to the spatiotemporal structure of our pure intuition via schemata, they result in the a priori principles of the understanding, in particular, the Analogies of Experience. It is in this way,

[29] The list of constituent concepts quoted in the previous note fills in the ellipsis.

[30] Here, then, is where Kant differs significantly with Newton over the role of metaphysics (compare note 6 above). That Kant has Newton in mind is confirmed by the closing paragraphs of the Preface, where Newton is named and cited (478–479). For Kant's "fabrication" and Newton's "feigning," see note 10 above. In the *hypotheses non fingo* passage Newton refers to "metaphysical" as well as "physical" hypotheses, thereby indicating that, for him, metaphysics involves hypotheses that have not yet been sufficiently supported by phenomena. Kant's point here is that metaphysics, in his sense, is not subject to this criticism.

[31] We saw in note 14 above, together with the paragraph to which it is appended, that Kant illustrates the a priori concepts of substance, causality, action, and force with the example of *moving forces* producing accelerations as alterations of state

as I have argued in Friedman (2013), that Kant thereby achieves a fruitful synthesis of Newtonian physics and Leibnizean metaphysics, where the latter has now been transformed into Kant's own revolutionary metaphysics of experience. However, just as physics has changed fundamentally since Newton, metaphysics or philosophy in the tradition initiated by Kant has also changed. So it is necessary to explore both sets of changes together in order fully to appreciate the legacy of Kant's achievement.

My own work in this direction begins from a neo-Kantian reinterpretation of the a priori in Reichenbach (1920), which distinguished between a priori in the sense of necessary unrevisability and a priori in the sense of constitutive of the object of (scientific) knowledge. The lesson of the theory of relativity, for Reichenbach, was that, while Kant himself had equated these two notions of the a priori, Einstein had shown that they must come apart. In particular, while Euclidean geometry and the Newtonian Laws of Motion are indeed constitutive, *relative* to this stage in the development of physical theory, the fundamental mechanical laws (characteristic of the Newtonian inertial frames) are changed in Einstein's special theory to those of relativistic mechanics (characteristic of the differently related inertial frames of special relativity). Moreover, even physical geometry is fundamentally changed in general relativity, so that we can presuppose in advance no particular (metrical) geometry at all, but only the more general structure of a four-dimensional (semi-)Riemannian manifold (of Lorentz signature).

My *Dynamics of Reason* (Friedman 2001) basically agreed with Reichenbach on this point, although some of the details varied. What is most important, however, is that I there added a historical account of parallel developments in scientific philosophy between Kant and Reichenbach, including such figures as Helmholtz, Mach, and Poincaré. In Friedman (2010), moreover, I discussed a number of additional contributions to the post-Kantian tradition, including such figures as Schelling, Cassirer, Husserl, and Weyl. The idea was better to appreciate, in both cases, exactly how such philosophical developments were intertwined with the scientific developments leading from Newton to Einstein and beyond – including some discussed above.[32]

It is precisely this intertwined set of developments, in my view, that provides the post-Kantian philosophical explanation of the "a priori

[32] Thus, e.g., Helmholtz was the main European thinker mediating between Maxwell's field-theoretic approach and the action-at-a-distance approaches favored on the continent; Hertz was Helmholtz's prize student.

sources" of the constitutive principles in question. These principles are a priori – in the relativized sense – rather than empirical, because they are what I have called enabling conditions for the more than inductive and more than hypothetical testing of properly empirical causal laws. But the "a priori sources" of these principles are provided by the corresponding developments in scientific philosophy, beginning with Kant's metaphysical foundation for Newtonian physics and continuing with the transformation of Kant's approach at the hands of Schelling, Helmholtz, Mach, Poincaré, and others.[33] Kant's metaphysics of experience is thereby relativized and historicized, along with the a priori itself.

[33] As suggested above, extending the Kantian notion of causal necessity also involves generalizing the notion of pure intuition, where experimental setups in which causal claims are tested take the place of Kant's forms of intuition; see note 26 above, together with the paragraph to which it is appended. This kind of consideration of the relationship between theoretical claims to be tested and the spatiotemporal properties of the corresponding experimental setups then takes the place of Kantian schemata. See Friedman 2012a for preliminary discussion; I hope to develop these ideas in more detail in future work.

CHAPTER 11

Metaphysical Foundations of Neoclassical Mechanics

Marius Stan

The most influential current accounts of Kant's philosophy of physics have it that our latter-day science of nature has shown his doctrine to be too narrow, thus no longer tenable in its original strong form. Its most glaring limitation is the geometric and chronometric structure – or determinate space and time content – to which Kant wedded it. Modern gravitation theory has shown that structure to be valid just in the classical limit. Thus, prominent interpreters have concluded, Kant's philosophical foundation must be read in light of the best science of his time, not ours – specifically, in dialogue with Newton's *Principia*, the high noon of early modern physics.

Suppose we are in the 'classical belt' of the world, where masses and speeds are low enough that relativistic theory is not needed. Has Kant's philosophy been refuted for that realm too? I claim that it has not. His views, I argue, remain as relevant as classical mechanics is alive, despite some premature reports of its demise in the wake of Einstein. To see my point, however, we must turn away from space or time structure and the Transcendental Aesthetic that legitimizes it. The most viable part of Kant's foundations, I suggest, is the transcendental *logic* – in its more determinate form as a constitution theory for objecthood in mechanics.[1]

Is that not a tall order? Will it not require me to try and defend anew the completeness of Kant's table of categories? No, not really. I am not asking here, what in the Transcendental Analytic can still explain the possibility of experience? Rather, I follow the Marburg neo-Kantians as I carve a new path alongside theirs. From Hermann Cohen to Michael Friedman, they have investigated how radical change of inertial structure affects Kantian

[1] Constitution theory is Kant's account of the "forms" that constitute the object of experience – specifically, the two forms of sensibility, and the *conceptual* structure supplied by the understanding. Here I am interested in the latter. I refer to Kant's *Metaphysical Foundations of Natural Science* simply as "*Foundations*" and cite it by volume and page number in the Academy Edition, from Kant 2004.

doctrine. I bracket their question entirely as I pursue my novel path. That is, I take classical structure – Galilean kinematics, Euclidean parallel transport, and absolute simultaneity – for granted, and focus instead on the *axiomatic* structure of mechanical theory. More exactly, I ask: what is the broadest genus of object for modern classical mechanics that we can anchor in Kant's categories and sensible forms? This question, I claim, has a cogent answer that shows Kant's metaphysics of material nature is viable and relevant, if we restrict it to its proper domain.

From my new vantage point, there is another ground for optimism. The conceptual basis of modern mechanics is an edifice with four parts: (1) fundamental kinematics, (2) a matter theory, (3) dynamical laws, and (4) a doctrine of objectivity. Kant had already anticipated this architecture, and called these parts Phoronomy, Dynamics, Mechanics, and Phenomenology, respectively. If we look at them from the perspective I have outlined, new insight emerges into his constitution theory of mechanical objects – and a new answer becomes possible to the question: do Kant's metaphysical foundations survive the challenge of modern science?

The question also bears directly on the central theme in this volume. We may regard Kant's constitutive principles in *Foundations* as laws broadly conceived: basic kinematic laws, laws of material constitution, general dynamical laws, and laws of mechanical objectivity. Then we can rephrase my question above as follows: are the laws of his "metaphysics of corporeal nature" still the laws of nature in a classical regime? Answer: yes.

But I must qualify my subject. Mechanics post-*Principia* is really the confluence of two broad streams of theory, viz., Newton–Euler dynamics and "analytic" mechanics. I set the latter entirely aside here. It poses serious challenges to my project, far from easy to solve, though I hope not intractable. I acknowledge them, and move on to examine Kant's chances of grounding the former. That holds out more promise and reward.

Still, the path to my conclusion is not apparent. Even circumscribed as Newton–Euler theory, modern mechanics rests on *two* distinct fundamental objects, viz., the mass-point and the deformable continuum.[2] Each object *by itself* supports a comprehensive theory of classical mechanics. And neither is explainable in terms of the other – they are mutually irreducible objects. Thus, from the stance I take here, we cannot answer my question above directly. So, I respond with a blend of reconstruction and

[2] There is a third unit of matter, the rigid body, which I leave out here. Kant argued resolutely against it, and it would be hard to brush that aside. And the rigid body is the preferred ontology of analytic mechanics, which I have set aside for now, as I explained above.

hermeneutics. I spell out modern constitutive assumptions for the two objects above and examine two respective strategies for embedding them in *Foundations.* Against that backdrop, I recommend the least disruptive path for the interpreter. Ipso facto that is also a case for the viability of Kant's metaphysical foundations.

Fortunately, Kant himself will help me in my enterprise. It is often thought that *Foundations* is a univocal account of a *single* kind of mechanical object. As I will show, that is not true. Kant speaks there in two voices, perhaps unwittingly, and he wavers between two species of object, discrete and continuous. This raises the question of which is really his considered view? Or, at least, which view is easier to defend on his behalf? My study thus is not just a case for Kant's continued relevance; it is also a fresh look at the deep structure of *Foundations*, which Kant obscures with his architectonic-driven mode of presentation.

However, his vacillation is hard to see without help from modern theory. Accordingly, I start with it, by spelling out material constitution, fundamental kinematics, dynamical laws, and objectivity criteria for a mass-point, the first kind of mechanical object (Section 11.1). And I do the same for the second kind, viz., the deformable continuum (Section 11.2). Then I turn to Kant and examine the resources for grounding either object in his constitution theory (Section 11.3). I end by assessing and recommending the most promising approach (Section 11.4).[3]

"Neoclassical" denotes a certain presentation of mechanics, more rigorous than the partial and often implicit "classical" version of Newton and Kant but less formal than the "post-classical" axiomatizations of Noll and his school.

11.1 The Mass-Point Approach

Kant's Dynamics is really a matter theory, and so its modern equivalent would be an account of material constitution. (I mean that as in physical theory, not intuition-based analytic metaphysics.) In current terms, one type of basic object, or unit of matter, is an entity the size of a point. This zero-sized object is endowed with two causally efficient, scalar-valued properties: inertial mass and charge (of various types). Mass grounds impenetrability in that no two masses can be superimposed, or made to occupy the same point-sized place simultaneously; it grounds resistance to

[3] Sections 11.1 and 11.2 are synoptic capsules that rely variously on Hamel 1909; Joos 1934; Malvern 1969; Truesdell 1991; Gurtin, Fried, and Anand 2010; and Wilson 2013.

acceleration by another mass-point, and also the ability to carry and transfer momentum. Charges give rise to scalar fields around the point, and they in turn induce fields of acceleration at every location for which they are defined. Some theorists capture these facts by saying that a mass-point mass has no true size, just an effective volume. On this view, macroscopic bodies are discrete ensembles, and mass-points are their ultimate actual parts.

Kinematics. An unconstrained mass-point has three degrees of freedom: three independent parameters specify its position completely. A motion is a change in degree of freedom; there is no intrinsic spin.

Because point masses are zero-sized and unconstrained *ex suppositione*, their basic kinematics is of quantities of straight-line motion, or time derivatives of their rectilinear displacements. The key grounding assumption – which Kant in Phoronomy grasped right away, and so we need to recover as we update his "phoronomic" foundations – is that these quantities are *additive*; more generally, they are linear combinations. (His preferred term for their additivity was "composition." Being directed magnitudes, he thought their additivity, ⊕, required explanation.) In light of that, a fundamental fact is that linear velocity $\mathbf{v}_i$ is resolvable:

$$\mathbf{v}_i = \mathbf{v}_x \oplus \mathbf{v}_y \oplus \mathbf{v}_z. \quad (1)$$

In words: it can be decomposed into, and recovered from, its projections, viz., component velocities relative to individual rigid coordinate axes *x*, *y*, and *z*. Another basic fact is that velocity of translation can be parallel-transported:[4]

$$\mathbf{v}_{B\ in\ \mathrm{F}} = \mathbf{v}_{\mathrm{F}\ relative\ to\ \mathrm{G}} \oplus \mathbf{v}_{B\ in\ \mathrm{G}}. \quad (2)$$

This is really an expression for the Galilean transformation, the framing principle of classical particle kinematics. It licenses redescriptions of motion across inertial frames of reference. A last basic principle is that linear acceleration too is additive:

$$\mathbf{a}_t = \mathbf{a}_1 \oplus \mathbf{a}_2 \oplus \ldots \oplus \mathbf{a}_n. \quad (3)$$

Namely, it is the vector sum of the single accelerations induced by individual mechanical agencies separately. In sum, the kinematic insights associated with this approach are three degrees of freedom, and composition, or vector addition (with resolution as its correlate).

[4] In the indices, "in" denotes velocity of mass points in respect to a frame, and "relative" denotes velocities of frames relative to one another.

Dynamics. Kant called "Mechanics" what we know as kinetics. Updated and tailored for the object of this approach, its "mechanical" properties are these. The full-fledged mass-point is endowed with two action-kinds, repulsion and attraction. Both are central forces in the weak sense; that is, they emanate from the point where the mass resides – because they are induced through charges carried by that point. Usually, these forces are the negative gradients of scalar-field potentials:

$$\mathbf{f} = -\nabla \mathbf{P}. \tag{4}$$

Just as usually, these potentials are irrotational. Then the forces are central in the strong sense: they point toward or away from the center, and their vectors lie on the straight line between the mass-points that exert and undergo them, respectively. Another way to describe these strong-central forces is by letting them be linear functions of some inverse power of the distance r between interacting particles[5]:

$$\mathbf{f} \ \alpha \ c \times 1/r^{\mathrm{k}}. \tag{5}$$

If a mass-point exerts both attraction and repulsion, and their respective force laws are species of **f** as in Eq. (5) above, it is easy to see that, on every radius from the mass-point, there is a point P at some distance b where the two forces balance each other: their respective accelerations cancel out. For irrotational potentials, the geometric locus of all these zero-acceleration points is a surface, viz., a 2-sphere with the mass-point at the center. This sphere is that which the young Kant in his lucid moments called the "sphere of activity" of a physical monad.

Beside laws of force-species as above, there are two dynamical laws of force *generaliter*. One is a version of Newton's *Lex Secunda*; more exactly, it is a conjunction of two statements:

$$\mathbf{f}_i = m \cdot \mathbf{a}_i. \tag{6}$$

In words, any force equals the mass of its target-point times the acceleration it *would* induce on the mass-point *if* it were the only force acting on it. And also,

$$\mathbf{f}_t = \mathbf{f}_1 \oplus \mathbf{f}_2 \oplus \ldots \oplus \mathbf{f}_n. \tag{7}$$

That is, impressed forces are linear-additive magnitudes. Often and tacitly, these two are stated together as the Second Law for mass-points:

[5] Below, "c" is a placeholder for algebraic combinations of dynamical factors – broadly speaking, types of charge and/or ratios of charge to mass. Uppercase boldface letters denote tensors, viz., 3×3 matrices, lowercase boldface vectors, and lowercase italics scalars.

$$\sum \mathbf{f}_i = m \cdot \mathbf{a}_t. \tag{8}$$

Namely, the resultant net force on a particle equals its mass times its effective acceleration. This ultimate dynamical principle of one-way action has a complement for interactions, the Third Law:

$$\mathbf{f}_{AB} = -\mathbf{f}_{BA}. \tag{9}$$

The force-vector on a mass-point *B* caused by a mass-point *A* is equal and opposite to the force-vector induced by *B* on *A*. In the standard case of mass-points with strong-central actions, the constraint (Eq. (5)) on force-laws entails[6]:

$$\left(\mathbf{r}_J - \mathbf{r}_K\right) \times \mathbf{f}_{JK} = 0. \tag{10}$$

That is, the impressed force by a mass-point *J* on another *K* is in the straight line between them. In turn, Eq. (10) together with the Eq. (9) make up a version suitably called the Strong Third Law: the mutual forces on two mass-points are equal, opposite, and *collinear.*

Objectivity. Finally, let us consider updating Kant's Phenomenology for mass-points. Their kinematic quantities take different values depending on the frame to which we refer them, so the question arises, which of these quantities are objective? Put differently: what features of motion are observer-independent?[7]

In our time there are two broadly overlapping approaches to objectivity for these mechanical objects. (1) Determine what quantities of motion are Galilean invariant, roughly, those having the same value for all the functions mapping a system of mass-points from one inertial frame into another while leaving their mechanical properties intact (McKinsey & Suppes 1955). From this standpoint, magnitude of linear acceleration counts as objective. (2) Establish the kinematic quantities that can be inferred (from initial conditions) by means of the dynamical laws. Notoriously, these laws entail Newton's Corollary V, which means that only acceleration is measurable by all observers. There is a complication here: this route allows certain unobservable linear accelerations.[8] With that proviso in mind, it is plain that both approaches single out the same quantity. Thus, we may

[6] The vector $(\mathbf{r}_J - \mathbf{r}_K)$ is the distance between the two mass-points *J* and *K*.

[7] I mean independent of *particular* observers. There is also a generic observer – the "transcendental subject," or "subjectivity" in general. In Kant's doctrine, *all* kinematic quantities are subjectivity dependent. But of course not all vary with particular subjects.

[8] By Newton's Corollary VI, a set of interacting mass-points may be subject to system-external forces that induce (at least within the boundaries of the system) a field of (nearly) equal and parallel acceleration, e.g., gravitational attraction by a very large body at a very large distance.

conclude, acceleration is objective – perhaps up to a common linear acceleration factor – but not velocity or position.

I move now to the next framework, to see what it requires. It too has four parts, which function as modern analogues of Kant's four disciplines in *Foundations*, adapted to continuous matter.

11.2 The Deformable Continuum Approach

The other paradigm of classical theory is continuum mechanics. Let us begin, again, with material constitution. To study extended bodies, continuum theory begins with an "Euler cut," a finite volume ΔV mentally carved out inside the body by a suitable intersection of planes or regular surfaces. There are two grounding assumptions. First, there exists a scalar function ρ, or mass density, defined everywhere in the cut, positive at every point; and also various charge density functions, also continuous everywhere. Second, any two material points in the cut can be moved relative to each other.[9] To handle this entity mathematically, we shrink it to an infinitesimal size as a volume element, dV, which has a true size. In this theory macroscopic bodies are continuous wholes, and volume elements are their least, potential parts.

Kinematics. Because continuous matter is extended and deformable, its geometric behavior is much more complex than any possible motion of single mass-points. The term "transplacement" is sometimes used for it, and it denotes a passing from a "reference" to a "current" configuration. A configuration is a mapping of a continuous body B into a Euclidean frame $\mathbf{E}^3$ at some instant t.

To stay with Kant's preferred language of "composition" in Phoronomy, a transplacement χ is any linear combination

$$\chi = \mathbf{F} \oplus \mathbf{t}. \tag{11}$$

Namely, a continuous-body motion is composed of a deformation **F** and a rigid translation **t**. Either can be zero, of course. Translations occur in mass-points, too, and are well understood. The topic of central interest in continuum mechanics is deformation, whose fundamental measure is the gradient **F**. To see what **F** means, start with the notion of a line element, or infinitesimal fiber, ds, that is, any material line segment whose end-points are two infinitesimally close points P and Q in the body. At any point M,

[9] The first constitutive assumption distinguishes continuous matter *radically* from mass-points; the second makes it radically distinct from rigid bodies.

the deformation gradient maps fibers (terminating in M) from the reference into the current configuration.[10] Globally, $\mathbf{F}$ informs about the change of shape and orientation of a continuous body. Locally, it tells us how fibers at M stretch, shrink, and change mutual angles, that is, rotate relative to each other. More exactly, let a curve s, parametrized by $\mathbf{z}(s)$, pass through M in the reference (undeformed) configuration, and let $\mathbf{x}(s)$ be the arc-length parameterization of the same curve in the deformed configuration. The deformation gradient $\mathbf{F}$ is the tensor of the partial derivatives of the "deformed" fiber $d\mathbf{x}$ with respect to the "undeformed" fiber $d\mathbf{z}$:

$$\mathbf{F} = \partial\mathbf{x}/\partial\mathbf{z}. \tag{12}$$

Another useful way to gain insight into $\mathbf{F}$ is to think of $d\mathbf{z}$ as the vectors tangent to all the curves s passing through M in the reference configuration, and of $d\mathbf{x}$ as the *same* tangent vectors in the current (deformed) configuration.

A basic idea in continuum kinematics is the polar decomposition:

$$\mathbf{F} = \mathbf{R}\cdot\mathbf{U}. \tag{13}$$

That is, a deformation can be seen as the product of a stretch tensor $\mathbf{U}$ and a rotation tensor $\mathbf{R}$. Stretching and shearing line elements around a point results in strain, the basic kinematic state induced by stresses and inducing stress in return. There are several gauges of strain; in elasticity and solid mechanics, a common measure is the (right) Cauchy–Green strain tensor $\mathbf{C}$[11]:

$$\mathbf{C} = \mathbf{F}^T\cdot\mathbf{F} = \mathbf{U}^2. \tag{14}$$

Thus strain can be gauged either from the deformation gradient at that point or just from the amount whereby line elements stretch or contract.

Dynamics. Mechanical continua exert two distinct kinds of actions. One is body force, a kinetic efficacy induced by a source directly on any point inside a continuous body; such is gravity or electromagnetism. Another type is contact stress, or traction. These act on the bounding surface of the body or the Euler cut, and are exerted indirectly – through the transmission of stress – on points inside the body.[12] Their effect is to

[10] If $\mathbf{F}$ has the same value at every point, the deformation is *homogeneous*, and inhomogeneous if $\mathbf{F}$ varies independently across the body.

[11] $\mathbf{F}^T$ is the transpose of the tensor $\mathbf{F}$.

[12] Strictly speaking, both types are really force densities (volumetric and areal, respectively). For each kind, values at a point are obtained by taking limits from values over volumes – or areas, for tractions – around that point.

stretch or compress lines, change angles, bend surfaces, and alter volumes, generally, to deform.

Dynamical laws must then take into account this duality of force. There are three principles governing the kinetic actions of continua. (Five, if we include thermal phenomena, which I leave out here.) One is Conservation of Mass, a conjunction of two theses. Here is one[13]:

$$m = m * . \tag{15A}$$

That is, the mass of a continuous body is invariant in all *Euclidean* frames. The other follows from it and the concept of deformation. It is a principle known as the Continuity Equation. I give the local form in the material description[14]:

$$\rho_0 = J\rho. \tag{15B}$$

The insight behind it is that, as mass remains constant throughout the stress action, only deformation can change mass density in a continuous body. The second fundamental principle is Cauchy's First Law, or the Balance Law of Linear Momentum. In essence it generalizes Newton's *Lex Secunda* in light of the dichotomy of force in continuous matter:

$$\rho a = \nabla \mathbf{T} \oplus \rho b. \tag{16}$$

Namely, the rate of change in linear momentum of the body equals the gradient of the external stresses plus the body forces acting on it. The last basic principle is Cauchy's Second Law, or Balance of Angular Momentum. It grounds the way in which efficient causes turn continuous volumes around axes of rotation:

$$\mathbf{r} \times \rho \mathbf{a} = \mathbf{r} \times \nabla \mathbf{T} \oplus \mathbf{r} \times \rho \mathbf{b}. \tag{17}$$

In words, the time-rate of change in angular momentum – around an arbitrary point at distance **r** from the body – equals the net torque of the body forces and contact stresses on a continuous body.

Objectivity. As the kinematic possibilities of continua far outnumber those of mass-points, the question of mechanical objectivity – the topic of an updated Kantian Phenomenology – for a continuous system poses new challenges. It really has two aspects. One regards the objective behavior of a special "representative" point in the body: its mass center, or any centroid

[13] Henceforth the asterisk denotes how an observer in rigid Euclidean motion relative to another sees or measures the same quantity as the first observer.

[14] ρ_0 is mass density in the reference configuration, ρ in the current (deformed) configuration, and J is the Jacobian of the deformation gradient tensor **F**.

more generally. This aspect considers the body as a particle, so the question has the same answer as it did for mass points – linear acceleration is objective, position and velocity are not. But this easy victory wins little; by reducing the body to the motion of its centroid, we leave unexplained most of material behavior, and turn extension and continuity into idle notions.

To avoid that, we must include true shape and stress. A different concept of objectivity arises then. To grasp it, start with the notion of change of Euclidean frame: let O and O^* be observers respectively at rest in two rigid frames in *arbitrary* relative motion. Let the following be functions of time: a point function χ, a symmetric tensor $\mathbf{T}$, a vector $\mathbf{c}$, and a proper orthogonal tensor $\mathbf{Q}$.[15] Let O observe χ and $\mathbf{T}$ at some time t. Then O^* will see these two magnitudes as χ^* and $\mathbf{T}^*$ given by:

$$\chi* = \mathbf{c} \oplus \mathbf{Q}{\cdot}\chi, \tag{18A}$$

$$\mathbf{T}* = \mathbf{Q}{\cdot}\mathbf{T}{\cdot}\mathbf{Q}^T. \tag{18B}$$

Intuitively, this says that O and O^* are in rigid motion relative to each other as they observe the same two functions.[16] Next, define a dynamical process $\boldsymbol{D}$ of a continuous body as a pair of fields $[\chi, \mathbf{T}]$ for a given assignment ρ of mass density, and let $\boldsymbol{D}^*$ be a process $[\chi^*, \mathbf{T}^*]$ for the same ρ. If their respective fields are related as in Eqs. (18a–b) above, the two processes are said to be equivalent. To say that $\boldsymbol{D}$ and $\boldsymbol{D}^*$ are equivalent means that the *same* physical process is described by two *different* arbitrary observers.

To complete the account, we need the idea of constitutive relations. Intuitively, they are differential equations – relating stress and strain – that describe possible basic ways in which continuous matter responds to applied stresses. An infinite number of such ways are possible mathematically, so – to delimit the class of physically admissible relations – an *objectivity* condition is imposed. This constraint is called "material frame-indifference." One way to state it is: "constitutive relations must be invariant under changes of reference frame" (Malvern 1969: 389). Namely, if a dynamical process $\boldsymbol{D}$ satisfies a constitutive relation F, then F must also be satisfied for any other equivalent process $\boldsymbol{D}^*$ as defined above. The crucial thing to note then is that a mechanical

[15] A tensor $\mathbf{Q}$ is properly orthogonal if the product $\mathbf{Q} \cdot \mathbf{Q}^T$ with its transpose equals 1.

[16] Because $\mathbf{c}$ is a rigid translation and $\mathbf{Q}$ a rigid rotation or mirror reflection. In other words, Eqs. (18a–b) are the Euclidean transformations.

theory of continuous matter needs two concepts of objectivity, or observer-independence: Galilean *and* Euclidean.

11.3 *Via Regia* and the Road Not Taken

It should be clear by now that the two approaches I outlined above are mutually irreducible, since their respective basic objects are radically distinct. One is discrete, zero-sized, exerting only Galilean-invariant actions at a distance. The other is continuous, extended, and capable of Euclidean-invariant contact actions. Kant's foundations cannot accommodate both objects. The reason is his grounding approach: the categories must be applied to matter – analyzed as mobile, impenetrable, momentum-carrying, observer-independent objecthood – so as to yield determinate *univocal* spatiotemporal content, viz., geometric magnitudes and their time derivatives. We do so by specifying the geometric makeup and kinematic content of the terms "body," "motion," "force," and "action." But a univocal specification yields a uniquely defined content, and so matter will be either mass-points or deformable continua, not both. So, we must decide which of these two is more firmly anchorable in his metaphysics.

One might object that my question has a foregone conclusion: Kant chose continuous matter explicitly. In *Foundations*, he claims to prove that mass distribution is continuous at all scales: "Matter is divisible to infinity, and in fact into parts such that each is matter in turn" (*MF* 4:503). And he argues strongly against "physical monads," that is, mass-points. So, it is clear which path to neoclassical mechanics is legitimate from his standpoint. Still, this is far from conclusive. For one, Kant does not much examine whether his preferred theory of matter coheres with the rest of his foundations, and the truth is that it does not (Stan 2014). In fact, a close look at his démarche in *Foundations* reveals that Kant vacillates between the two approaches I have outlined, though he seems unaware of that. Thus, my question stands. Then let us examine both approaches and assess their respective chances and interpretive costs.

Mass-points. Grounding their kinematics is uncommonly easy, for Kant has done so already, though unintentionally. In "Metaphysical Foundations of Phoronomy," he starts with mobility as an essential property of body and goes on to justify the key principle for the "composition of motions," viz., an addition rule for *linear* velocity, which he regards as the basic quantity. But he legitimizes that conception with a sleight of hand:

> Since in phoronomy nothing is to be at issue except motion, no other property is here ascribed to the subject of motion, namely, matter, aside from movability. *It can itself so far, therefore, also be considered as a point*, and one abstracts in phoronomy from all inner constitution ... If the expression "body" should nevertheless sometimes be used here, this is only to *anticipate* to some extent *the application of the principles* of phoronomy to the more determinate concepts of matter that are still to follow. (*MF* 4:480; emphasis added)

Satis non est. Point motion is just change in three degrees of kinematic freedom. Kant's doctrine is that bodies are continua, but those are inherently extended and have shapes irreducible to points. Ergo, Phoronomy properly conceived will be a theory of change in shape. At the very least, Kant ought to have made it a theory of infinitesimal fibers *ds* and *their* relative motions. Then the basic concept of Phoronomy becomes deformation, *not* point motion, as he claims. His claim above does not follow from his argument but is licensed by two other considerations. (A) In some contexts, we may abstract from the body's rotations and deformations and treat it as a mass-point, by letting the point-motion of a centroid stand for the motion of the extended body itself. That is precisely Kant's practice in *Foundations*, where his preferred centroid is the mass center – because he aimed to ground Newton's approach, who in *Principia* replaces any extended planet with an (extensionless) mass-point. (B) Or we may regard the body as *composed* of mass-points, and then *its* motion is the sum of its constituents' point-motions.

Be that as it may, the fundamental kinematics for the approach (A) is very much the basis for (B) too, and so Kant's mistaken conclusion above becomes his *felix culpa*. It led him to a composition theorem equivalent to Eq. (1) above. To see that, let a test particle move simultaneously relative to three frames themselves in mutual uniform translation at right angles. The particle's three motions – relative to each frame – can now be treated as *its* component velocities, and so its "composite motion," as Kant would call it, becomes its linear velocity $\mathbf{v}$ as in Eq. (1), relative to a (fourth) frame supposed stationary. Thus in the course of explaining how motion becomes a magnitude, Kant justified the fundamental result in particle motion. Even better, if we regard the motions in Kant's account as velocities acquired from rest by a particle, it turns out that his explication of "composite motion" can also justify Eq. (3) above. And because Kant took "composite motion" to be at bottom an addition of interframe velocities, naturally his notion also explains the Galilean transformation (Eq. (2)) above. Consequently, Kant has already grounded the basic kinematics for mass-points, and the interpreter need only reembed it in a broader explanatory framework.

The dynamics stage of my inquiry must follow two tracks. We may wonder first about the chances of embedding in his doctrine the two force species – themselves generic – proper to mass-points. Here too the official Kant is already of great help. A constant of his natural philosophy was that "force" is constitutive of matter.[17] One such agency, "attraction," is just the kind of force that mass-points exert too: central, induced by a potential, acting at a distance. Kant conceived it on the model of gravity but was noncommittal enough to leave its specifics open. Thus he would agree that the law of his "attraction" takes the general form as in Eq. (5) above. However, his other dynamical agency, "repulsion," defies the interpreter, for in the 1780s he made it a contact force, thus incompatible with the mass-point approach. I have two remarks to help overcome this obstacle. First, Kant's account of repulsion is not self-standing. It is a corollary of his broader view that matter is a physical continuum: "divisible to infinity . . . into parts such that each is matter in turn," he argues in Proposition 4 of Dynamics (*MF* 4:503). But his master argument for continuous matter in that Proposition is not unassailable; if we examine it critically, it may turn out to be flawed.[18] Second, even if his argument is good, we can discount it as part of the price for extending his foundations to make room for modern mechanics. It is a small price, really: keep his force of repulsion, discard his continuity-of-mass thesis, and the result is repulsive force fit for mass-points; just like his attraction, it would be central, acting at a distance. Then its force-law will be a species of Eq. (5) above. So, it can balance Kant's fundamental "attraction," thereby cogrounding a notion of least effective volume as mass-points have. After all, mass-points used to be Kant's theory of matter in *Physical Monadology*.[19] And so, dropping Proposition 4 gets the interpreter a ready-made basis for half of basic particle mechanics.

What about the other half, the general laws of impressed force? Here too the landscape is friendly. Kant in his Phoronomy gives grounding for Eq. (3), viz., the additivity of linear acceleration. He offers that as the "a priori basis" for another key result, the Parallelogram of Forces. Though he does not carry out an actual derivation – alluding instead to the collinearity

[17] He meant that "force" is a basic attribute – with mobility – in that bodily extension, impenetrability, and ability to transfer momentum are explainable from "force."

[18] For instance, Friedman – virtually alone in having examined it – seems to me hesitant to declare it sound (2013: 150f.). I take comfort in that.

[19] Or one of them – apparently, he flirted with two theories in that paper; cf. Smith 2013.

of force and its induced acceleration – Kant is sure that it can be done, and so are some interpreters.[20] If so, Kant has thus secured Eq. (7), the additivity of impressed force. Still, that is not yet enough for the Second Law. We must still identify Kantian resources for Eq. (6), the *Lex Secunda* for single forces. Perhaps an argument can be made – broadly in tune with the Kant of Friedman's recent work – that Eq. (6) is not really a law but an axiom: a postulate, or physical primitive, posited for the sake of rigorous experience, or, as Friedman has it, for making possible the application of mathematics to nature, by exhibiting how force becomes a magnitude, that is, acquires a ratio structure (2013: 237). So, we have prima facie reasons to be hopeful about extending to mass-points this part of Kant's foundations, though of course it needs more work.[21]

Discarding his Proposition 4 as I urged above would also make the Strong Third Law at home in Kant's metaphysics of nature. He already has a good argument for Eq. (9), the weak version of *Lex Tertia.* With attraction and repulsion reconceived as *actiones in distans*, it is easy to show these Kantian basic powers are strongly central. We need rely only on an idea already available to him, specifically Lagrange's Ω-function (Lagrange 1777). The function is a mathematical description of the basic agency exerted by mass-points: a scalar potential whose gradient is a conservative, irrotational vector field. In physical terms, if attraction and repulsion are Ω -functions, they obey Eq. (4); just relabel "P" as "Ω" to see that. But Eqs. (4) and (9) entail (10), which means that Kant's mature metaphysics – with physical monads, the direct consequence of deleting Proposition 4 – support the Strong Third Law, hence the two basic laws of point–mass dynamics.

And so we come to objectivity and the task of extending Kant's Phenomenology. The prolegomenon to a solution, I believe, is to note a deep ambiguity in his theory of motion. On the one hand, modern interpretations (based on Friedman's highly consilient reading) strongly imply that Kant lets true motion be the kinematic quantity satisfying his "mechanical laws." This idea coupled now with an ontology of mass-points entails that true motion is that which arbitrary inertial observers would

[20] See Friedman 2013: 377–379 and Stan 2014: 429. Still, we should give this matter careful scrutiny. Attempts to ground the Parallelogram of Forces are notoriously thorny, and it is not yet clear that, even by eighteenth-century standards of physical proof, Kant's derivation would succeed.

[21] Additionally, we would have to supplement Kant's conception of *mass*. In the 1780s, he thinks of it just in terms of active powers – of transferring momentum and "imparting motion" by acceleration fields – but hardly as the measure of *resistance* to acceleration. That is unfortunate, especially as in *Physical Monadology* Kant had a clear conception of mass as resistance to force.

measure by applying the laws (6) through (10). Specifically, it is linear *acceleration*, because those laws are Galilean relative. On the other hand, Kant throughout his philosophical maturity was wedded to the early modern idea that the laws of motion are "true relative to" a preferred frame, viz., absolute space. However, this makes true motion –thus objective behavior – be a preferred *velocity*, or rate of change in distance to the privileged frame. That runs against the first strand of thought above, and also against our modern view of objectivity.

That is the conundrum. To break it, I recommend adopting the modern view that Kant lets the laws themselves define objective mechanical behavior, and adjust the Phenomenology – construed à la Friedman – to the dynamical laws of point masses. As far as I can tell in advance, the only drastic correction will involve Kant's doctrine of necessary motion, admittedly the acme of Phenomenology. Officially, Kant construes it as true velocity, whose necessity is allegedly conceptual: it "follows *immediately* and unavoidably from the *concept* of the relation of the moved in space to anything else movable thereby" (4:558, emphasis added).[22] That becomes untenable in the project pursued here. Like Friedman, we ought to reread "necessary motion" as true acceleration, whose necessity is special-metaphysical, or grounded in Kant's "special metaphysics of material nature." The Second Law, now added explicitly to that edifice, entails that any force on a body, or set of mass-points, induces an objective acceleration on it. *Lex Tertia* then makes it *necessary* that any body experiences an objective acceleration in proportion to its action on other bodies, because all impressed forces necessarily are interactions, by the Third Analogy.[23]

Continuous matter. Let us now move on to the other paradigm and assess it for ease of retrofit to Kant's metaphysical foundations. It seems we should be hopeful about grounding its fundamental kinematics. Recall that a continuum motion *generaliter* is composed of a rigid translation **t**, a rigid rotation **R**, and a pure stretch **U**. And Kant sees his task in Phoronomy precisely as supplying a priori principles to exhibit the "concept of a composite motion" in pure intuition (*MF* 4:487). So, we might expect the core grounding to be in place already. But this hope is short lived. Kant sees composition of motions as homogeneous in two senses: it is a uniform operation, vector addition, and is a "composition of the homogeneous" in

[22] He means that, just by a conceptual analysis of "relative motion," we can establish that two interacting bodies have true velocities relative to each other, in the mass-center frame.

[23] It is not clear what role is left for Kant's absolute space to play. The dynamical laws pick out a family of equivalent frames, not a single preferred reference system.

that it adds up motions of the same kind (*MF* 4:495; cf. also Sutherland 2014). But continuum motion is *heterogeneous* on both counts. It involves vector addition and tensor multiplication; see Eqs. (11) and (13) above. And it combines different species of displacements. More seriously, Kant's basic kinematic object is of the wrong kind for continua. His relevant entity is "the motion of a point" (*MF* 4:489). But in continuum kinematics the fundamental object is the infinitesimal *fiber*, not the point, so any interpretive attempt to anchor continuum motion in his metaphysical foundations of phoronomy is bound to violate the letter of Kant's doctrine. Perhaps we can retain its spirit, at least? For instance, by discarding wholesale Kant's outdated geometric approach to kinematics and replacing it with "composition" as in linear algebra, that is, as combination of elements in a vector space. I cannot judge the feasibility of this proposal in advance, but I must warn that ultimately it will need anchoring in Kant's categories of quantity, as his foundational program requires. Thus, so far continua seem a good deal harder to embed in his Phoronomy.

Next, it appears that updating his Dynamics, or theory of matter, is easiest. After all, he does have the continuity of mass and a distinction between "penetrating" and "surface" actions, that is, body forces and stresses (*MF* 4:516). And yet even here some reconceptualization is needed. Kant really ought to start with a notion of force *density* – specifically, the amount of body force per unit volume and of stress per unit area. Values at a point, needed for equations of motion, are then obtained by taking limits, viz., letting these units shrink to a volume element dV and area element dA, respectively. Thus, not even this chapter of *Foundations* would survive intact the evolution of subsequent mechanics.

On to the dynamical laws, then. Kant saw insightfully that a mechanics of continua needs Conservation of Mass, which he linked firmly to his metaphysics through the First Analogy of Experience. But his architectonic approach to grounding yields gains *and* losses. If we follow his lead, we can at best generalize his procedure to cover inertia of rotation and the equality of impressed torques. However, it is far from clear how the interpreter might make room for principles that really drive theory-building in this mechanics, viz., Cauchy's two laws (16) and (17) above. The latter principle is probably not out of bounds for Kant: it is derivable as a theorem, provided we allow on Kant's behalf that the stress tensor $\mathbf{T}$, the basic measure of contact action, is symmetric. Still, that raises again, more pressingly than ever, the question of whether the Second Law for single forces is anywhere in Kant's metaphysical foundations, and how in its absence he might be credited with it.

I think there is a heuristic out of this predicament, and it appears feasible. Friedman once ascribed to Kant the view that *Lex Secunda* was implicit in his "mechanical" laws of inertia and of action–reaction (1992a: 143). I suggest reviving that idea in the context of rational reconstruction rather than textual hermeneutics.[24] More generally, I think we may regard the Second Law (and its correlate for rotation) as "bridge laws" between the purely metaphysical, qualitative principles of force and action, and the mathematically redescribed "mechanical" appearances, that is, phenomena of motion conceived as transplacements. This would keep the spirit while correcting the letter of Kant's *Foundations* as an explanatory project of how mathematics can apply to appearance and turn it into mechanical experience, paradigmatically.

Still, I am hard pressed to see any Kantian path to grounding objectivity for continua. Granted, Kant apparently had a relativity principle compatible with Newton's Corollary Six. However, even that is too weak for the objectivity at issue here, which is really Euclidean relativity, as it were: invariance across rigid frames, translating *and* rotating relate to each other. But Corollary Six holds only in a proper subset of the equivalence class of all Euclidean frames. Nor is there any hint of how this notion might be reconciled with Kant's categories of modality. Because continuum objectivity does not single out inertial frames, the interpreter cannot just transfer the stepwise approximating procedure outlined in the Phenomenology – as Friedman and others read it – and invoke the modal category appropriate for each stage.[25] Continuum objectivity is a constraint meant to determine "real" or "intrinsic" *material* properties – codified by the body's response to stress, not just by the motion of its mass center, which Kant singled out for treatment in Phenomenology. It is the presence and special nature of stress and strain, which mass-points lack and cannot support, that requires a second, distinct concept of objectivity for continuous matter.

11.4 Balance and Conclusions

In sum, to ground continuum theory in Kant's metaphysics of nature, we would need to discard his Phoronomy and replace it wholesale; retool the mathematics of his Dynamics – and generally replace his geometric

[24] There is textual support for my suggestion. Kant regards his "dynamical" repulsion and attraction as forces of "imparting motion" whereby they generate momentum increments. And his indirect proof of the Equality of Action and Reaction assumes that actions are impressed forces that generate linear accelerations (*MF* 4:536; 562f.). These claims presuppose the Second Law.

[25] Of course, inertial frames remain indispensable – the laws of continuum mechanics hold just relative to them, not generally.

methods with the partial-differential calculus; supplement his Mechanics with two highly nontrivial principles; and regard his Phenomenology as incomplete, in need of extension by another concept of objectivity, material frame-indifference.

In turn, adopting mass-points would leave the Phoronomy mostly unaltered; leave the Dynamics in place, save for the untenable Definition 4; require us to make explicit the source and status of the Second Law in his Mechanics; and prompt us to clarify his idea of true motion in Phenomenology.

With the initial estimate of interpretive costs behind us, which of the two approaches is preferable? On balance, mass-points appear a good deal less radical and invasive. Continuum theory requires strong foundations, and I am really not sure that Kant's metaphysics can provide them or even that it needs to do so.

Another point against the continuum approach is that Kant's project pursues a stringent notion of metaphysical grounding, and so it places strong constraints on any attempt to extend or update it. Specifically, explanation in it flows from the categories and transcendental principles through the concept of matter and its attributes to basic mechanical theory. But keeping continuous matter makes it very hard to live up to this ideal, hence give a *Kantian* grounding for neoclassical mechanics. That is because, if we keep Kant's physical continuum, we must supply what he lacks – everything else, really. In effect, we would have to build fundamental kinematics and dynamics, the results (11) through (18) above, from the ground up, that is, from Kant's analysis of matter – and then retrofit them to his constitutive apparatus imported from the first *Critique*. How to carry this out is utterly unclear to me, yet it must be done or else we betray Kant's foundational thrust, which renders hollow any victory so obtained. In contrast, mass-points satisfy Kant's strong constraint very well. He had already shown – accidentally, as he rejected them in theory – that they instantiate his categories of quantity, relation, and modality. It is quite easy now to argue that they cohere with the categories of quality too: they bear the same basic forces as his continuum, and the continuity of mass distribution has *no* import on category-instantiation.

Here is another reason to extend Kant in line with the first approach. Note how little disruptive it is to anchor mass-point mechanics in his doctrine. That is because the doctrine is really a palimpsest. The architectonic of categories gives it a monolithic façade, but behind it two different modes of thought, particulate and continuous, are superimposed. Thanks to Smith (2013), now we know that Kant always wavered between these two views of matter. Officially a mass-point theorist in *Physical Monadology*, Kant there at times lets the continuum picture take over unannounced. In *Foundations*, he commits to continuous matter, and yet he cannot let go of the particulate approach specific to mass-points. Then, seeing as his foundation is really a heterogeneous blend, the interpreter should not be afraid to purify it by removing extraneous ingredients. In my view, taking the mass-point approach amounts to proper purification: it retains the essence of Kant's philosophy of mechanics while increasing its strength. In contrast, the continuum approach would dilute it beyond recognition merely to preserve just one commitment out of so many that Kant expressed in *Foundations*.

A last but compelling reason to embed mass-points into Kant's metaphysics is that they make continuum theory *foundationally* dispensable. This is not immediately apparent, but emerges in hindsight from the long-term growth of mechanics. Modeling matter as continuous is needed just at certain scales, not universally, and mass-points have proven resilient enough to allow that. Specifically, modern mathematical tools, calibrated for the desired scale, enable mechanics to start with discrete mass-points and define *from* them continuous fields as needed.[26] So, it turns out that mass-points are the better matter theory for classical dynamics, and the physical continuum is a local, reducible phenomenon. Then we need not saddle Kant with it as we bring him into the present.

I end with a parting note, not unimportant. Readers will wonder how my key thesis – that we ought to commit Kant to a kind of physical monads – is compatible with his Second Antinomy and its injunction against them. My answer is brief here, but it deserves elaboration elsewhere. In two respects, my result is orthogonal to Kant's concerns in the Antinomy. For one, I do not advance it as a thesis in transcendental realism, viz., about the architecture of matter as *Ding an sich*. On the contrary: it is about the microstructure of the substratum, itself *phenomenal*, of all appearances in outer sense. It recommends an ontology inferable from phenomena – of motions and deformations given at mesoscopic scales and above – by a systematic use of Kant's Postulates of Empirical Thought. Then my result is in line with his broader project in the Analytic, just as I promised at the outset. Because of that (now this is another respect) my reasoning is safe from Kant's censure in the Antinomies. I did not reach my conclusions by mere analysis from mere

[26] The first effort along these lines was Irving and Kirkwood 1950; the latest, most detailed expression is Murdoch 2012.

concepts, aided by the Principle of Sufficient Reason – as Wolff and the practitioners of "rational cosmology" did, thus earning Kant's rebuke. Rather, I ascribed to Kant mass-points as the best ontological explanation of all determinate phenomena of motion in a Galilean regime. As such, my inference is beyond Kantian reproach, for it does not claim for itself any greater certainty than he believed he had secured for his own principles in *Foundations*.

In conclusion, there is a promising case that Kant's "special metaphysics of material nature" remains viable and relevant. As I have explained, that is due to his insightful grounding of mechanics in a robust, resilient constitution theory of mechanical objecthood and to the changing but ultimately happy fortunes of classical mechanics itself. In regard to the former, Kant's far-sighted kinematic and dynamical foundations permit an ontology of mass-points to be embedded without much disruption – though not the continua he embraced in the Critical decade. As to the latter, while Einstein did curtail the once imperial ambitions of mechanics, in the end a vast swath of determinate experience still unfolds at speeds and scales for which classical theory remains indispensable. Thus, thanks to his relentless and deep engagement with it, Kant's mechanical foundations remain the most important legacy of early modern natural philosophy.

Acknowledgments

For helpful suggestions and enlightening discussions, I am indebted to Michela Massimi, Katherine Brading, Angela Breitenbach, and Michael Friedman.

PART V

Laws in Biology

CHAPTER 12

Laws in Biology and the Unity of Nature

Angela Breitenbach

12.1 Introduction

Kant's philosophy of science is famous for putting the lawful unity of nature center-stage. Kant argues that all natural phenomena are law-governed. Any appearance of lawlessness is only the result of our ignorance, and this is true whether we consider the animate or inanimate world. Kant puts this point unambiguously at the start of the *Jäsche Logic*:

> Everything in nature, both in the lifeless and in the living world, takes place according to rules, although we are not always acquainted with these rules. – Water falls according to laws of gravity, and in the case of animals locomotion also takes place according to rules. The fish in the water, the bird in the air, move according to rules. The whole of nature in general is really nothing but a connection of appearances according to rules; and there is no absence of rules anywhere. If we believe we have found such a thing, then in this case we can only say that we are not acquainted with the rules. (*JL* 9:11)

Kant here uses "rules" in the place of "laws," but his central claim is clear.[1] There is no exception to the law-governed character of natural phenomena. Kant develops this claim in the *Critique of Pure Reason* where he sets out the fundamental laws of nature in general including, for example, the principle that every event has a cause. In the *Metaphysical Foundations of Natural Science*, he furthermore presents the laws of specifically material nature such as the rule that every change of motion requires an external cause. The arguments Kant develops for the universal reach of these laws in the first *Critique* and the *Metaphysical Foundations* are complicated but the basic idea is well known. The necessity these laws express is not given independently in things in themselves but is grounded in principles of the human intellect.

[1] Kant sometimes employs the term "rules" (*Regeln*) in contrast with "laws" (*Gesetze*) to designate not empirical laws of nature but empirical regularities, e.g. at A646f./B674f., or regulative principles, e.g., at A645/B673. In other passages, however, he uses the terms interchangeably, e.g., at A189ff./B234ff. For the purposes of this chapter, I shall treat the two terms as equivalent.

It is human understanding that prescribes these laws to nature, and this is why we can know a priori that all of nature in general is unified by causal laws and all corporeal nature by the laws of motion.

Kant is also concerned to account for the more specific, empirical laws of nature, such as Newton's law of universal gravitation, Bernoulli's principle of fluid dynamics, or Hauksbee's law of gases. He elaborates on this dimension of the lawful unity of nature in the Appendix to the Transcendental Dialectic, the introductions to the *Critique of Judgment*, and the logic lectures. On Kant's account, the necessary form of empirical laws is grounded in the a priori principles of understanding, but cognition of their empirical content requires reflecting on the specificities of individual phenomena.[2] Kant recognizes that such reflection consists in comparing and contrasting particulars, following methods of induction and analogy, and that such empirical research presupposes the use of reason and its regulative idea of systematic unity.[3] More specifically, Kant argues that in searching for empirical laws we must assume that our cognitions of the diversity of natural phenomena can be organized under systematically related concepts and principles. We must presuppose that all particular cognitions, no matter how seemingly unrelated, can be understood as part of "a system connected according to necessary laws" (A646/B676).[4] As Kant argues further, this regulative idea of the unity of all cognitions, or the unity of science, presupposes an idea of the unity of the objects of such cognitions, or the unity of nature.[5] In our search for particular laws we must assume that nature is such that it can indeed be explained by a unified science. We must regard nature as itself a system of law-governed phenomena hierarchically ordered into lower-level "species" and higher-order "genera" (A651f./B679f.).

The lawful unity of nature thus lies at the center of Kant's philosophy of science. We can know it as constituted by the fundamental a priori laws that determine the universal character of all natural phenomena. And we must furthermore think it as ordered into a hierarchy of kinds and governed by systematically interrelated empirical laws.[6]

[2] I argue for this reading of Kant's conception of empirical laws in Breitenbach (forthcoming).

[3] See, e.g., the discussion at *JL* 9:132f.

[4] The nature and viability of this presupposition is addressed, with different conclusions, in the chapters included in Part II of this volume.

[5] Kant also presents this move from the principle of the unity of science to that of the unity of nature as the move from a logical to a transcendental principle. See A648ff./B676ff.

[6] Kant accordingly distinguishes these two conceptions of unity as the (constitutive) "distributive unity" of the understanding and the (regulative) "collective unity" of reason (A644/B672). I assume both when I refer to the unity of nature in this chapter.

A second feature for which Kant's philosophy of science is known is the specific diversity and autonomy Kant grants the individual sciences. In particular, he holds, there is an important difference between the study of physical and biological processes. Kant is acquainted with the work of the early German biologists, including Johann Friedrich Blumenbach and his students. Kant is aware of their experiments with polyps studying the regeneration of amputated parts.[7] Against this background, he rejects mechanistic accounts of organic nature, denying that organisms are explicable merely according to mechanical laws. In the *Critique of Judgment* he draws the famous and equally remarkable conclusion:

> It is quite certain that we cannot adequately come to know the organized beings and their internal possibility in accordance with merely mechanical principles of nature, let alone explain them; and indeed this is so certain that we can boldly say that it would be absurd for humans even to make such an attempt or to hope that there may yet arise a Newton who will make comprehensible even the generation of a blade of grass according to natural laws that no intention has ordered; rather, we must absolutely deny this insight to human beings. (*CJ* 5:400)

Kant suggests that the progress made in physics by Newton's formulation of the laws of motion could never be achieved by the discovery of equivalent laws in biology. Organisms, he argues, cannot be explained by the laws of matter but must instead be judged according to a regulative maxim of purposiveness. Biological entities are special, on this account, since they require consideration under teleological principles.

Kant thus argues, on the one hand, for the systematic and law-governed unity of nature, and, on the other, for the autonomy and mutual irreducibility of the concepts that characterize the domain of different sciences such as physics and biology. A key question this raises is how these two claims go together. If organisms cannot be explained according to the laws that unify all material natural phenomena, then what are the implications for Kant's conception of the lawful unity of nature? Should we reconceive this unity in light of Kant's philosophy of biology?

A number of commentators have recently concluded that Kant's teleological conception of biological entities and processes poses a challenge to the idea that organisms form part of the lawful unity of nature. They have argued that the only way to avoid the disunity of nature,

[7] See, e.g., *CJ* §§80–81.

for Kant, is to deny that biological entities are strictly speaking part of nature. On these accounts, since the principles particular to our conception of organisms are purely regulative, they are unable to permit knowledge of biological phenomena, let alone of biological laws.[8]

By contrast with these views, my aim in this chapter is to show that Kant's teleological construal of the organism is compatible with a naturalistic conception of biological entities and, in particular, leaves room for the discovery of genuine biological laws. Rather than taking organisms out of nature, I argue that Kant's teleological conception is a means for reflecting on parts of nature as organic, thereby picking out natural phenomena to be studied by the biologist. Moreover, I suggest that, on Kant's account, construing organisms according to teleological principles does not threaten but, instead, sheds important light on the unity of nature. Kant's notion of natural teleology, I argue, offers a model for thinking about the unity of nature in a way the supposedly paradigmatic examples of the physical sciences do not.

To develop these claims, I begin by setting out my reading of Kant's teleological conception of the organism. I suggest that this conception is grounded in an analogy with the end-directed causality with which we are familiar from our own reason (Section 12.2). I then show that this reading leaves room for the possibility that organic phenomena are governed by fully naturalistic biological laws and form part of the systematic order of nature (Section 12.3). I argue, furthermore, that the possibility of naturalistic laws in the life sciences leaves the lawful unity of nature intact without eliminating the need for a teleological construal of the organism. However far the biological sciences progress, teleological principles remain indispensable and inform our search for a unified conception of nature (Section 12.4). The two aspects for which Kant's philosophy of science is famous should thus be understood in conjunction. Kant's idea of the unity of nature leaves room for, and indeed requires, a diversity of mutually irreducible concepts and laws (Section 12.5).

[8] According to Zammito 2012: 123, it remains "a philosophical conundrum" for Kant how organisms can be "integrated into a unified system of empirical laws as the 'order of nature.'" Similarly, Richards 2000: 26 concludes that Kant's *Critique of Judgment* "delivered up a profound indictment of any biological discipline attempting to become a science." See also Beiser 2006. I agree with these authors that Kant's teleological conception of organisms precludes an endorsement of Blumenbach's *Bildungstrieb* as an empirical teleological principle, thus shedding doubt on Lenoir's 1980 claim that Kant's conception of teleology significantly shaped the evolution of the life sciences in the late eighteenth and early nineteenth centuries. I disagree with the further claim, however, that for Kant "biology could never be a science at all" (Zammito 2012: 122).

12.2 Organisms as Natural Purposes

Organisms appear to pose a problem for the lawful unity of nature on Kant's account. As Kant argues, they are contingent according to the laws of matter. In the Critique of Teleological Judgment, he maintains that "nature, considered as a mere mechanism, could have formed itself in a thousand different ways without hitting precisely upon the unity" we find in organisms (*CJ* 5:360). Kant elucidates the characteristic unity of organic beings, which accounts for their mechanical inexplicability, in three important steps. First, he argues that mechanical laws not only underdetermine the specific unity of particular organisms but that these laws are in principle incapable of accounting for the form and functioning that distinguish organic from inorganic nature. Second, Kant maintains that we can make sense of the distinguishing features organisms only according to teleological principles. To conceive of organisms as lawful, we must regard them according to principles of purposiveness. However, as Kant claims in a third point, teleological considerations are not explanatory. They have analogical status and function only as a regulative guide in the search for explanations. Teleological laws characterize our reflection on organisms, but they are not constitutive of nature itself. To shed light on this teleological conception of organic beings, I consider these claims in further detail below.

First, Kant recognizes with Blumenbach that organisms have not only a special organization but also a capacity for self-organization. The form and functioning of the organism's parts depend on the form and functioning of the organism as a whole. Moreover, the maintenance of the whole is in turn ensured by the way in which the parts bring about and sustain each other. The structure and movement of the bird's wings, for example, are determined by the placement of the wings within the animal body and by their functional contribution to the bird's capacity to fly. At the same time, the movement of the wings is dependent on nutrition received from blood circulating through the animal body, a system that also has the function of maintaining the body's health and regenerating damaged organs. Organisms, as Kant argues, are not like machines but are internally self-organizing and striving for their own existence and survival. As he concludes, "an organized product of nature is that in which everything is an end and reciprocally a means as well" (*CJ* 5:376).

Organisms thus construed are contingent in a strong sense. Their contingency is not exhausted by their being underdetermined by actual conditions in combination with the laws of matter. With respect to such underdetermination, organisms are no different from any other empirical phenomena, for on Kant's account the particularity of individual phenomena is never exhaustively explained by general laws. As Kant argues, however, organisms are contingent in a special sense, since mechanical laws cannot even in principle explain the species of organization and self-organization characteristic of plants and animals. According to the laws of matter, material wholes are functions of their parts and of the way in which those parts interact and form into wholes. But reducing an organic whole to its parts and to their mechanical interaction can never explain the dependence of parts on the whole and their characteristic interdependence within the whole.[9] Actual conditions together with the laws of matter are thus incapable of accounting even for the general features of the form and functioning of organisms.

The only way to make sense of organic beings as law-governed, Kant therefore argues in a second step, is to think of them as the products of purposive activity and as themselves exhibiting goal-directedness. As he puts it, "the concept of the combinations and forms of nature in accordance with ends is still at least one more principle for bringing its appearances under rules where the laws of causality about the mere mechanism of nature do not suffice" (*CJ* 5:360). Once we construe the organism as purposive, we can see the form and functioning of its parts as law-governed. We can regard them as standing under teleological laws, that is, under rules for the realization of a purpose, in this case, the organism.

Consider, for example, the inner working of the bird's body. According to teleological principles, it is governed by such rules as "in order to maintain the proper functioning of the wings, the limbs must receive nutrition through the blood," or "for the purpose of maintaining blood flow, the heart must perform a steady pumping function." If we thus think of an intention as having arranged the parts so as to bring about the whole, and if we regard the working of the parts as means for bringing about the proper functioning of the whole, we can recognize organic beings as law-governed.[10] Their lawfulness does not consist in their being necessitated by the laws of matter. Instead, it is constituted by their being necessitated by practical rules whose employment is a means for the realization of an end. It is this purposiveness of organic beings that Kant

[9] For a detailed account of the mechanical inexplicability of organisms in Kant, see Breitenbach 2006; cf. McLaughlin 1990: 152ff. and Ginsborg 2015: 281ff.

[10] Kant thus presupposes a general conception of law that covers both theoretical and practical principles. See Watkins 2014a and Chapter 1 in this volume.

calls the "lawfulness of the contingent" (*CJ* 5:404). And since organisms are part of nature, he characterizes them as "natural purposes" (*CJ* 5:361). The reason why organisms are not explicable in mechanical terms, on this reading, is that we always, implicitly, regard them as natural purposes whose form and functioning is teleologically and not mechanistically structured. And in order to make sense of such a purposive unity, we therefore have to regard it, explicitly, as governed by teleological principles.[11]

At this juncture it is important to bear in mind that, for Kant, purposes are essentially tied to intentionality and rational agency. On Kant's account, a purpose is the aim of goal-directed activity. At times, Kant defines purposes as the *realized* ends of intentional activity, as "that the concept of which can be regarded as the ground of the possibility of the object itself" (*CJ* 5:227). At other times, he characterizes purposes as the *conceived* aims of goal-directed activity, as "the concept of an object insofar as it at the same time contains the ground of the reality of this object" (*CJ* 5:180). Since, as I have suggested, Kant's organic analogy involves the purposive organization of a realized end as well as the self-organizing activity of acting for a conceived end, both notions are implied in Kant's conception of natural purposes. Most important, and more generally, purposes "have a direct relationship to reason" (*TP* 8:182), and the capacity to set ends can therefore be found only in rational beings. However, Kant is also convinced that nonhuman nature, including non-human organisms, is essentially nonrational and that by implication non-human nature has no intentions. This leads him to the difficulty of having to account for the apparent "contradiction" in the very notion of a "natural purpose" (*CJ* 5:370).

Kant resolves the tension by arguing, in a third step, that we do not cognize or experience organic nature as purposive in the ordinary sense, but rather "project" (*CJ* 5:360) the concept of a purpose onto nature. The concept, he suggests, has a purely regulative function. As a nonnatural concept, it guides our thinking about and our inquiry into biological entities and processes, even though it cannot feature in any determining judgments about such entities or processes. More specifically, Kant suggests that our judgments about organisms as natural purposes have analogical form. In considering organisms as natural purposes we regard

[11] In Breitenbach 2014 I have argued that the implicit and explicit employments of teleological principles are the two functions of teleological judgment.

them "in accordance with a remote analogy with our own causality in accordance with ends" (*CJ* 5:375).

This "remote analogy" does not consist, as a common reading has it, in the relation between organism and artifact.[12] For, as Kant points out, artifacts are in important respects distinct from organisms. Most important, artifacts do not display the capacity for self-organization that Kant regards as a key feature of organic beings. Kant's remote analogy is rather that "with our own causality in accordance with ends," that is, with the intentional activity that is productive of artifacts and, in particular, with the intentional activity of reason itself. According to this analogy, we consider the parts of an organism as if they were arranged so as to perform a function within the whole, and we regard organisms as themselves containing the principle for the proper arrangement and functioning of their parts.

The teleological analogy thus identifies the way we reflect on organisms and their striving activity with the relations between reason and the purposive actions of reason. More precisely, the analogy concerns two sets of relations.[13] First, it identifies the relation between organisms and their striving activity with the relation between reason and its goal-directed actions. Second, it identifies the relation between organisms and their parts with the relation between reason as a whole and different rational capacities. In other words, the teleological analogy suggests that organic beings are purposively directed toward their own ends just as we set ourselves ends and strive for their realization. And it indicates furthermore that, just as our rational activities are purposively related to realizing and maintaining our rational agency, so the working of the parts of an organism is purposively related to ensuring the existence and survival of the organism as a whole. In regarding organic beings as natural purposes, we thus implicitly reflect on them by means of this two-fold analogy with the capacity for end-directed causality with which we are familiar from our own reason.

In response to the problem that organisms seem to pose for the lawful unity of nature, Kant has thus put forward a teleological account.

[12] For the common reading, see, e.g., McLaughlin 1990: 37ff. and Guyer 2005: 352ff. Ginsborg 2015: 281ff. points out the difficulties with the artifact analogy, but argues that the difference between organism and artifact is not important for understanding the teleological character of living beings. For a detailed discussion of Kant's analogy, see Breitenbach 2009a: chapters 3 and 4.

[13] Kant characterizes analogies as postulating the identity of two relations, i.e., as presenting not "an incomplete similarity between two things, but rather a complete similarity between two relations of wholly dissimilar things" (*Prol* 4:357; see also A179/B222).

Although organic beings, construed as natural purposes, are inexplicable by the laws of matter, they are lawful if construed in accordance with teleological principles. As Kant is keen to stress, this teleological construal has regulative status; it can guide analogical reflection on organic nature, but it cannot ground objective cognition of organisms.

Kant's answer to the problem organisms seem to pose now raises our original questions with new urgency. We were concerned to understand how the supposedly contingent organic phenomena could form part of the law-governed unity of nature. If Kant's solution to this problem relies on an analogical conception of organic beings, however, how can it ground cognition of the laws that govern those beings? Moreover, if organisms are construed as lawful merely by analogy, how does Kant's account solve the problem of biological entities seeming to lie outside the unity of nature? I put forward an answer to these questions in the next section.

12.3 Laws in Biology

We can begin answering the problem of biological laws by appreciating how Kant's analogical conception of the organism relates natural phenomena to a nonnatural idea. On the reading I propose, the analogy with rational purposiveness does not offer grounds for any constitutive claims about organisms, but it makes possible reflection on, and identification of, particular natural phenomena. The important implication of this is that the analogy can guide research into the phenomena thus identified and, specifically, into the laws that govern these phenomena. The answer I propose in this section is thus that Kant's teleological conception leaves room for a naturalistic account of biological laws, which situates biological entities within the systematic order of nature.

As we have seen, it is an important feature of Kant's teleological analogy that it cannot lead us to determinately apply the concept of purposiveness to organic phenomena. As Kant puts it, what is transferred from one side of the analogy to the other is only "the form of the reflection, not the content" (*CJ* 5:351). In other words, we reflect on organisms as if they have the internal structure and directedness of rational purposiveness, yet we do not thereby assert that a principle of reason is active in nature. However, analogical reflection of this kind nevertheless guides the formation of concepts and principles that do determine organic phenomena. One way to understand this guiding function is to see that, in thinking about organisms by analogy with the intentional activity of reason, the nonnatural idea of such an activity provides an analogue for the missing empirical concept. As Kant argues, teleological analogies come into play because the laws of mechanical causality are insufficient for subsuming the experienced organic phenomena under "rules" (*CJ* 5:360). In the absence of a suitable naturalistic concept, the super-sensible idea of purposiveness guides the way we pick out natural phenomena as unified objects, which can also be understood in naturalistic terms and in accordance with the laws of nature.[14]

What this entails, more concretely, is that, by means of the teleological analogy, we reflect on different spatial and temporal parts of the organism as belonging to one unified being. For example, we regard the rabbit's nose, teeth, and intestines as related to the organism as a whole just as means are related to an end. Moreover, we reflect on the animal's acute sense of smell when it detects some but not other weeds as edible, and the complex working of the animal's intestines when they digest these weeds, as functioning for the purpose of the animal's survival. In judging natural entities and processes in this way, we thus regard them as unified by the idea of an organized and self-organizing whole. It is in this sense that the analogical reflection presents a condition for identifying organic beings and picking them out from their environments. In considering the rabbit, by analogy, as standing under teleological laws, we are not taking the organism out of nature, judging it to be something nonnatural. Instead, we are using a nonnatural idea to guide reflection on a natural phenomenon. The important insight of this Kantian proposal is that judging organisms in this teleological manner is a means for reflecting on parts of nature as unified natural bodies.

An implication of this is that the natural phenomena thus picked out can in principle be studied also in naturalistic terms. Indeed, if organisms are to count as natural beings "one is required to pursue [naturalistic explanation in causal-mechanical terms] as far as one can" (*CJ* 5:388). As Kant puts it,

> the mere teleological ground of such a being is . . . inadequate for considering and judging it as a product of nature unless the mechanism of the latter is associated with the former, as if it were the tool of an intentionally acting cause to whose ends nature is subordinated, even in its mechanical laws. (*CJ* 5:422)

[14] In line with this, Kant draws a parallel between schemata and analogical presentations as performing analogous functions (see *CJ* 5:351).

Kant's teleological account of organisms thus requires that the parts of nature identified by means of the teleological analogy stand under the universal laws of matter. Even though these laws by themselves do not explain organisms as natural purposes, they nevertheless govern the parts of nature on which we reflect by means of the analogy. By asking for the purpose of a biological trait, for example, we can thus direct our attention to the causal-mechanical processes that determine the trait's causal role within the organic body. The teleological analogy can thus guide research into natural processes to be explained by mechanical laws.

One may wonder, of course, whether the laws of nature to be discovered in this way are restricted to the physical-mechanical laws. Or does Kant's teleological account of organic nature leave room for specifically biological laws beyond the universal laws of matter? Turning, for a moment, to the more recent literature, it is interesting to see that philosophers of science have raised a parallel question about the possibility of laws in the life sciences today. In response to this question, several authors have suggested that the only laws that exist in biology are physical and, perhaps, chemical laws.[15] Their reasons have been various, but high among them is the view that natural evolution is a process full of contingency and that, for this reason, neither the process nor the entities it produces are characterized by necessity relations. According to this view, no distinctly biological laws are there to be discovered. Or, in other words, the only laws to be discovered would not account for the specific contingency and exceptional character of organic nature. Such laws would not be specifically biological laws. Instead, they would be concerned with the exceptionless regularities of physical or chemical processes.

A similar problem may also be raised for Kant. As I have argued, Kant's teleological conception of organisms makes room for research into non-teleological laws. But once we turn our focus on the natural phenomena picked out by means of the teleological analogy, and consider these phenomena independently of any teleological principles, we should not expect to discover laws that explain the characteristic organic properties we ordinarily regard in teleological terms. It thus seems that explanations by means of entirely naturalistic laws would not so much explain, but rather explain away, distinctively organic features.

Initial appearances to the contrary, however, I believe that Kant's account leaves room for the possibility of biological laws. Such laws would have to fulfill two desiderata. First, in order to be genuine laws of nature,

[15] E.g., Rosenberg 1994 and Beatty 1995. For critical replies, see Sober 1997 and Mitchell 2000.

they would have to be thoroughly naturalistic; that is, they would have to make use exclusively of concepts that determinately apply to natural phenomena. They would have to employ causal, nonteleological concepts. Second, in order to qualify as specifically biological laws, they would have to employ some specifically biological concepts. Such concepts would have to be suitably naturalistic, too. They would present naturalistic counterparts, for example, to the teleological notion of an organism or the related teleological concepts of a species or eco-system that we find in Kant.[16]

Consider, for example, recent attempts in the philosophy of biology to define the concept of an organism. There is little agreement among contemporary biologists and philosophers of biology about which organism concept to employ, with proposals invoking, for example, such diverse criteria as spatial boundaries, life-cycles, genetics, fitness maximization, or metabolic autonomy.[17] All parties to the disagreement, however, share the conviction that the concept of the organism can be cashed out in naturalistic terms. In particular, and important for our case, there is growing agreement that the organism concept has to be regarded as basic and irreducible for the purpose of certain inquiries. It has been suggested, for example, that we cannot understand evolutionary processes unless we focus on the organism, and not only the gene, as a fundamental explanatory concept. And it has been argued, in a similar vein, that understanding how molecular parts interact is dependent on understanding how they are organized in a system, where the organism is recognized as a key level of organization.[18] In biology and the philosophy of biology, we thus have no scarcity of candidate concepts that are entirely naturalistic and, at the same time, regarded as basic for certain parts of biological inquiry. It is concepts such as these, I suggest, that would feature in genuine biological laws and explanations by means of these laws.

Kant is not, of course, aware of any such concepts nor of any biological laws that employ these concepts. And yet he is careful enough to remind us that we cannot determine in advance the limits of naturalistic explanations.[19] His account thus leaves room for the possibility of examining the specific properties of the natural phenomena picked out as organisms, and of classifying them and generalizing over them. We can, for example,

[16] For Kant's teleological discussion of species and races, see his *ODR*, *HR*, and *TP*. For his discussion of what today we call "eco-systems," see *CJ* §§63 and 82.

[17] Clarke 2013: 415 counts fifteen different proposals. See also Clarke 2010 and Wilson and Barker 2013. Similar disagreement surrounds the species concept. See, e.g., Ereshefski 2001 and Okasha 2002. On the eco-system concept, see Sterelny 2001.

[18] See Nicholson 2014 and Walsh 2015.

[19] See *CJ* 5:388 and 418.

examine the traits of their organs and the relations between their parts, and find out that rabbits have an excellent sense of smell and that their digestion works by hindgut fermentation.[20] We can distinguish between rabbits and hares and discover that they have common ancestors and belong to the taxonomic order of Lagomorpha. We can formulate higher-level generalizations such as Mendel's law in order to explain the process of inheritance in rabbits and other animals. In asking about the purpose of a particular trait, we can thus inquire into its causal function within the animal system or its contribution to the evolutionary advantage of the organism's ancestors. In this way, we can research into the features that characterize the natural phenomena we have picked out as organisms and the associated laws that govern them.[21]

Kant's account is thus compatible with the possibility that genuine biological laws may be discovered by future scientists. To be sure, such laws could never explain the teleological character of organisms as natural purposes. Instead, they would explain the entirely natural, and hence causal, properties of those parts of nature on which we ordinarily reflect by means of the teleological analogy. In doing so, moreover, they may go some way toward explaining why biological entities and processes appear to us as suitable for reflections in accordance with the teleological analogy.

Organisms construed naturalistically in this way do not pose any challenge to the lawful unity of nature. They do not threaten unity under the fundamental a priori laws and, in particular, offer no counterexample to the universal reach of the laws of matter. Nor are organisms thus construed at odds with ambitions for a systematic understanding of nature according to specific empirical laws. For it is in principle possible to spell out how the naturalistic biological concepts on which such laws rely relate to higher-order chemical and physical concepts and how the biological laws are in turn systematically related to chemical and physical ones. Organisms thus construed naturalistically are determined, just as other

[20] Hindgut fermentation is the fascinating process by which the digestive organs ferment hard plant tissues that the animal would otherwise be unable to digest.

[21] In Breitenbach 2009b, I have furthermore argued that Kant's teleological conception is compatible with the discovery of causal relations that may consist, e.g., in the natural selection of a particular trait, according to evolutionary explanations, or in the trait's nonlinear causal contribution to a system, in accordance with systems theory. Since neither of these theoretical accounts alleviates the need for teleological reflection, I have argued that neither offers what would be required of the Newton of the grass blade that Kant deemed impossible. In this I agree with a conclusion reached by Cohen (forthcoming). I think it is also important to recognize, however, that Kant's account leaves room for explanation in terms of biological laws even though such explanation can never ultimately dispose of the need for teleology.

natural phenomena are, by specific laws – in this case, biological laws – and can be shown to be part of the systematic unity of empirical nature.

And yet this conclusion raises the question of whether the teleological perspective can eventually be eliminated from biology. As I set out in the previous section, Kant argues that organisms are contingent according to the causal-mechanistic laws of matter. Does the possibility of fully naturalistic concepts and laws in biology imply that we must give up on Kant's contingency claim? Can Kant's analogical conception of organisms as natural purposes eventually be reduced to the nonteleological conceptions just considered? These are the questions I turn to next.

12.4 The Organic Unity of Nature

Kant makes a strong claim in response to the question of whether we might, in the end, replace the teleological account of organisms with exclusively naturalistic concepts. He maintains that however far we proceed in our causal explanations, any such explanation would nevertheless require subordination under teleological laws. As he puts it, any mechanical explanation of organic beings would "always be subordinated to a teleological principle as well" (*CJ* 5:417).

One way to begin making sense of Kant's thought is to see that teleological reflection characterizes the way in which we ordinarily make sense of the world. It constitutes our prescientific, everyday understanding of nature. In other words, we cannot give up reflecting on nature according to teleological principles, since such reflection fundamentally characterizes what it means for us to live in a world that is, at least in part, alive. By projecting thoughts that we associate with our own rational purposiveness onto natural phenomena, we reflect on those phenomena in a way that would not be possible without the teleological analogy. By means of this analogical reflection, we first make sense of parts of nature as organisms.[22]

The fact that, as Kant suggests, we are the kinds of creatures that construe the world as consisting of organic and inorganic things, does not of course tell us why we do so or whether we might not learn to do otherwise. And it may simply be the case that as a matter of fact we are wired so as to pick out organisms among other natural phenomena. Against this claim to philosophical bedrock, however, I believe that a longer and more comprehensive story can be told on Kant's behalf. The details of the story are not always made fully explicit by Kant and can only

[22] I develop this answer in more detail in Breitenbach 2008.

be hinted at here. The key idea can be garnered from his writings on teleology and from the overall project of a "critique of judgment as a means for combining" (*CJ* 5:176) theoretical with practical philosophy and as "the mediating concept between the concepts of nature and the concept of freedom" (*CJ* 5:196). According to this idea, the necessity to construe nature in teleological terms is ultimately grounded in our need to understand the relation between reason and nature. More specifically, however far we proceed in replacing teleological considerations with causal-mechanistic explanations in biology, we cannot but think of ourselves in teleological terms, as acting intentionally and aiming at ends. Understanding ourselves as beings that act for purposes and realize ends in the natural world, moreover, in turn requires reflecting on nature, including organic nature, in accordance with teleological principles. On Kant's account, I thus suggest, the need to understand ourselves as rational agents in the natural world makes the teleological conception of nature indispensable to us.

To shed some light on this proposal, imagine the biologist's attempt to replace any teleological conception of organic nature with naturalistic concepts and principles. Reducing teleological reflection to causal explanations in this way would make it apparent that the conception of organisms as natural purposes, with which we started, was merely an analogical projection. It would show that our teleological reflections on organic nature make a claim not about the biological phenomena but only about ourselves. As Kant puts it,

> The concept of a thing as in itself a natural purpose is therefore not a constitutive concept of the understanding or of reason, but it can still be a regulative concept for the reflecting power of judgment, for guiding research into objects of this kind ... ; not, of course, for the sake of the cognition [*Kenntnis*] of nature or of its original ground, but rather of the very same practical faculty of reason in us in analogy with which we consider the cause of that purposiveness. (*CJ* 5:375; my translation)

Kant argues that teleological reflection does not give us determinate insight into the nature of organisms, but it can throw light on the character of our own reason. It can represent to us the purposively ordered and goal-directed activity of our own rational capacities, reflected in nature. While we cannot cognize nature as a natural purpose, we can thus find out something about ourselves by considering nature in teleological terms.

Even construed as merely regulative and analogical, the conception of organisms as natural purposes thus relies on a teleological conception of ourselves. Moreover, Kant believes that, as agents in the natural world, we necessarily think of ourselves as acting for purposes. We must regard ourselves as acting intentionally, in particular, when we study the natural world including organic entities and processes. Our reductive ambitions in biology thus have to stop at our own case. Even if we aim to replace teleological reflection with causal-mechanistic explanations of organisms, we cannot eliminate reference to purposes altogether. Moreover, once we have admitted this much teleology into the picture we cannot end here but must also make room for teleological reflection on nature as a whole, including biological phenomena. This is because, in thinking of ourselves as setting ends and acting for purposes, we need a grip on what it means for our rational purposes to be effective in nature. And we cannot make sense of ourselves as acting and realizing ends in nature, according to Kant, unless we also regard nature as itself "a teleological order of things" (*CJ* 5:379).

Two points are important for making sense of this step from the teleology of our own reason to a teleological construal of nature as a whole. The first concerns the epistemic status of the required story about reason in nature. On Kant's account, any such story will be purely speculative. Since we cannot achieve determinate understanding of purposes in nature, any account of the relation of reason and nature will be merely regulative. The second point concerns the details of this relation. Kant thinks that we can construe reason in nature only if we assume that nature itself is purposive for the development of reason. And this, in turn, requires that we assume nature has a teleological ground and is, as a result, ordered in its entirety according to relations of purposiveness. The attempt to make sense of ourselves as acting in the natural world, on Kant's account, thus relies not only on a teleological conception of our own rational activities but also, importantly, on extending teleological reflection to nature as a whole. It presupposes the idea of nature as a system in which everything is related as means and ends and thus contains organic entities and processes that maintain and sustain each other. The idea of the whole of nature as purposive for the development and expression of reason thus brings with it the conception of organic beings as natural purposes.[23]

Kant elaborates on this speculative account in the *Critique of Judgment* and in other writings on teleology. In *Idea for a Universal History with a Cosmopolitan Aim*, he spells out a teleological "guiding thread" for a history

[23] For a more detailed discussion of the teleological conception of nature as a whole, see Breitenbach 2009a: chapter 6. Watkins 2014b argues that Kant's conception of nature as a system of ends is related to the totalizing urge of the faculty of reason, which extends purposive relations to all of nature.

of the realization of reason in nature (*Idea* 8:30). In *Conjectural Beginning of Human History*, moreover, he offers a teleological "conjecture" for construing the origin of reason in nature (*CB* 8:109). In the third *Critique*, finally, he spells out the idea of the whole of nature as a "system of ends," purposively ordered and directed at the existence and realization of reason (*CJ* 5:377). Interestingly, here it is the organism as natural purpose that provides the model for this teleological conception of nature as a whole. As Kant puts it,

> by means of the example that nature gives in its organic products, one is justified, indeed called upon to expect nothing in nature and its laws but what is purposive in the whole. (*CJ* 5:379)

According to Kant, organisms as natural purposes offer an example for extending teleological consideration to all of nature and, thereby, for representing nature as purposive for human reason. Kant thus construes the unity of nature on the model of the organism, projecting the idea of a purposive system with its own end and its own organizing principle onto that of nature as a whole. On this model, reason, as the final end, is conceived as organically developing out of nature, while ultimately pointing beyond nature itself.

The need to make sense of our own dual nature as rational beings that act in the natural world thus ultimately makes the teleological conception of nature indispensable for us. Even if in the life sciences we aim to reduce considerations according to teleological principles to explanations by means of causal laws, we will nonetheless have to retain the teleological conception of a natural purpose. All progress in the natural sciences notwithstanding, on Kant's account, our teleological conception of the organism remains irreducible.

I suggest that this indispensability of teleological considerations of nature has crucial implications for Kant's account of the unity of nature. Two results are particularly important. First, as we have just seen, the teleological conception of organisms as natural purposes helps us elucidate the idea of the unity of nature. In the introduction I sketched this idea as that of a system of kinds, hierarchically ordered into genera and species. Kant's proposal in the third *Critique* implies that we can shed light on this indeterminate idea by means of the model of organic unity. On Kant's account, organisms function as an analogue, or a symbol, of nature as a whole.

The second implication is that the irreducibly teleological conception of organisms provides us with a reason for picking out living beings as a basic focus of investigation and explanation. As we saw above, according to Kant's idea of the unity of nature, we must presuppose that all natural phenomena are systematically ordered in a hierarchy of genera and species. On this picture, no particular kind of phenomenon is in any way privileged, or has importance over any other. Every phenomenon can be explained, albeit incompletely, by laws on a more general level.[24] The need for a teleological conception of organic nature, however, gives us reason in biology to prioritize one of these levels over others. In other words, because of the indispensable need to consider organisms according to irreducible teleological principles we have reason not to reduce the organism concept to anything below or above the organic order. The teleological conception thus guides us in the search for laws and explanations that take the organism concept as basic.

In this general sense, then, the teleological conception of the organism would guide the study of living beings even if we had a fully naturalistic organism concept. It carves out a part of nature as an object of study in its own right. It highlights a level of organization in the hierarchy of kinds on which to focus scientific investigations.[25] Kant's conception of natural teleology as an irreducible heuristic is thus compatible with, and indeed sheds important light on, his idea of the unity of nature. Kant's regulative account of teleology informs his idea of nature as a lawful unity that leaves room for different sciences with their distinctive concepts and principles.

12.5 Concluding Remarks

As I have argued in this chapter, Kant's philosophy of science contains two ideas that, prima facie, stand in tension with each other. On the one hand, Kant promotes the idea of the lawful unity of nature. According to this idea, all natural phenomena stand under universal a priori laws and are ordered in a hierarchy of systematically related empirical concepts and laws. On the other hand, Kant also puts forward the thought that there is a diversity of irreducible concepts and principles, which carve out the domain of distinct sciences and guide our inquiry into particular parts of nature. According to this second idea, different concepts and principles are necessary for different types of scientific inquiry – in particular,

[24] Rauscher 2010: 295 makes this point clearly when he argues that in the systematic unity of concepts in accordance with which we understand the unity of nature "there is no privileged point, with the single exception of a possible highest universal concept at the top."

[25] Similarly, Quarfood 2004 has suggested, though for different reasons, that teleological judgment provides biology with its object.

regulative teleological principles for biological research. The combination of these ideas raised the question of this chapter. How could phenomena that require such irreducibly distinct principles belong to one and the same law-governed whole?

In response to this question I have argued that Kant's two ideas are compatible and that understanding them in conjunction sheds important light on the idea of the lawful unity of nature. As I have suggested, the indispensability of irreducible teleological principles in biology is grounded in a central fact about ourselves, in our need to make sense of the possibility of reason in nature. Only if we leave room for principles of purposiveness can we make sense of nature as including ourselves as rational and purposively acting beings aiming to understand the natural world. Furthermore, as regulative principles, teleological laws not only are compatible with the unity of nature but, crucially, function as means to identify the object of biological inquiry. They guide research into particular kinds of natural phenomena and the biological laws that govern them.

On the reading of Kant I have put forward, organic entities and processes are thus part of the lawful unity of nature, and biological laws are possible. Far from falling out of nature and the domain of the natural sciences, as some readers have concluded, there can be a genuine science of organic phenomena and of the laws that govern them. We can hold on to Kant's conception of the lawful unity of nature, while making room for a plurality of irreducible concepts and laws.

Acknowledgments

Earlier versions of this chapter were presented at the final conference of the Leverhulme International Network on *Kant and the Laws of Nature* held at the University of Edinburgh and at conferences at the Freie Universität Berlin, TU Dortmund, and the Swedish Collegium for Advanced Study. I am grateful to the participants for their helpful feedback. I owe special thanks to Yoon Choi, Alix Cohen, Michela Massimi, Tiago Mata, Sasha Mudd, and Eric Watkins for constructive comments and discussions. Last but not least, I am grateful to the Leverhulme Trust and the Riksbankens Jubileumsfond for supporting my research.

CHAPTER 13

The Building Forces of Nature and Kant's Teleology of the Living

Catherine Wilson

Nature, according to the Comte de Buffon's influential formulation, is "that system of laws established by the Creator for the existence of things and for the succession of beings" (1764, XII:iij). It is

> a living power, immense, embracing everything, animating everything ... at once cause and effect, mode and substance, design and work; altogether different from the productions of human art, which are only dead works, Nature is a work perpetually alive ... time, space, and matter are her means, the Universe her object, movement and life her purpose. (ibid.)

But what were the laws of life? The success of Newton and the Newtonians in systematizing terrestrial and celestial mechanics could hardly be extended to the animal and vegetable kingdoms. The living world, unlike the stately system of stars and spherical planets separated by vast stretches of void, was a tangled and dense profusion. It appeared to be structured, but in two entirely different ways: hierarchically, as a ladder of being extending from mosses and fungi at the bottom to human beings and perhaps angels with celestial bodies at the top, and horizontally, according to the old models of genera and species.

The Linnean system of classification made evident but gave no clue as to the significance of the diversity of life forms, from worms and mollusks to quadrupeds, nor to that of the existence of a basic vertebrate skeletal structure, or to the unnerving similarity of the orangutan and man. Buffon comments that where living beings are concerned, "it appears as if Nature has thrown out all at once a world of related and unrelated beings, an infinity of harmonious and inharmonious combinations, and an endless cycle of destructions and renewals" (1749 I:11). Hume refers to "blind nature, impregnated by a great vivifying principle, pouring forth from her lap her maimed and abortive children, without discernment or parental care" (1778/1998: 74). The laws of life seemed to be only the inevitabilities of sex, reproduction, and death.

Kant's interest in the science of living nature led him to address a number of separate though related problems within this intellectual context, including the ultimate origins of life in the distant history of the planet, animal generation in established species, and their extinction and transformation. In addition, his queries extended to the existential and metaphysical question of the purpose, if any, of the passing parade of generations and the significance of the extinctions of the past. In this chapter, I set out his thinking on these topics in relation to his underlying commitment to the requisites of any science of nature and his innovative theology. However averse Kant was to the doctrine of hexameral creation, which he refers to as "poetic raving" (*CJ* 5:410), he was even more averse to the notions that human beings were chance productions.[1]

It is commonly held that Kant maintained a strict and productive distinction between a "constitutive" mechanical science based on the Newtonian forces of nature and a "regulative" teleological approach. In this chapter I argue that this is not quite right. Kant maintained a "constitutive" belief in active, organizing forces resident in matter throughout his career. In his early writings, he regarded these forces as lawful, "blind," and unproblematic. His earliest commitment to teleology consisted in his notion that the universe had an overall purpose: namely, the production of intelligent beings in a display of divine magnificence in nature's sheer productive power to create worlds and world systems on a vast spatiotemporal scale. Where these morphological forces were concerned, however, he quickly reached the conclusion that they could not have formed the *first* plants and animals from ordinary matter, but only their successors, via the transmission of the seminal material in which they inhered.

When Kant returned to the problem of life in 1790, his interest in teleology had shifted. He was now deeply puzzled as to how blind forces could construct and maintain such a peculiar entity as a plant or animal, whose purpose was to live and whose parts and mechanisms were all conducive to that end. The morphological force or forces involved were clearly different from attraction and other moving forces, though they were somewhat similar to the chemical forces that could produce such curious structures as crystals. He concluded that they, unlike all the other forces known to science, had to be conceived on analogy with goal-driven, intelligent action. Their reality and efficacy was never in question for him.

[1] Translations of *CJ* are from Immanuel Kant. 1987. *Critique of Judgment*. Translated by Werner S. Pluhar. Hackett.

An account of Kant's intellectual development over fifty years or so between 1754 and his death in 1804 properly begins with his adaptation of the famous naturalist Buffon's theory of the Earth in his own *Allgemeine Naturgeschichte*.

13.1 Kant and the Universal History of Nature

Buffon's *Histoire Naturelle*, whose German translation began to appear in 1751 with a Preface by Albrecht Haller, was one of the most popular publications of the eighteenth century, and the young Kant drew heavily on its first volumes as a lecturer in physical geography. The aim of the earlier work, in addition to delighting, entertaining, and serving as an encyclopedic survey of the animal kingdom, was to advance the cause of materialism. In its many volumes and supplements, published between 1749 and 1779, Buffon referred time and again to "the Creator," but in such a way as to indicate that he meant only to refer to the productive powers of nature. That the planet Earth had evolved to its present state over hundreds of thousands or even billions of years was a thesis increasingly supported by geological evidence.[2] And in the first volume of the *Histoire*, Buffon explained – by exclusively physical principles – the formation of the planets, the Earth's history, and the changes in its geological and topological features, beginning with its birth as a flaming lump torn from the sun by a passing comet and including the universal deluge. He ridiculed the notion of the English cosmologists that the divine will and the sinfulness and needs of man were in any way implicated in the Earth's history and features.[3]

The young Kant approved and imitated this project, drawing as well on other published sources (Jaki 1981: 13–26). His aim in the 1755 *Universal Natural History and Theory of the Heavens: An Essay on the Constitution and Mechanical Origin of the Entire World-Edifice Treated According to Newtonian Principles* was to show how matter alone, once the laws of attraction and repulsion were impressed onto it, could gradually bring worlds and "world-orders" (*Ordnungen*) – solar systems, nebulae, and the Universe as a whole – into existence without direct agency on the part of God. He rejected the notion that any celestial features of the world-system or meteorological features of our planet were created for our use or for any

[2] Leibniz's *Protogaea* 1749 circulated in manuscript from 1693 and was an early contribution to this genre. For a full account, see Rudwick 2014.

[3] Further background is given in Roger 1997: 93–115.

other purpose and that every region in nature is full of life, and essentially good. "The very same unlimited fruitfulness of [Nature] has brought forth the inhabited celestial globes as well as the comets, the useful mountains and the harmful crags, the habitable lands and the empty deserts, the virtue and the vice" (*UNH* 1:347; J 181).[4] At the same time, the process was purposive, insofar as the plenitude of habitable planets that were produced expressed the infinity of God's nature, and insofar as the process generated a succession of ever more brilliant beings able to comprehend and contemplate the glorious spectacle intellectually (*UNH* 1:352; J 184) and perhaps optically; Kant's awe in the face of the "starry heavens" (cf. *CPrR* 5:161) was lifelong.

Never again did Kant refer as confidently and frequently to the productive powers of the "general laws of nature" (*allgemeine Gesetze, Wirkungsgesetze*) to fashion the visible world as in the *Natural History* and in the critical discussion of physico-theology in his 1763 essay, *The Only Possible Argument in Support of a Demonstration of the Existence of a God.* Published at first anonymously in 1755 with a dedication to Frederick II of Prussia, the *History* was reportedly reprinted under Kant's own name in the following year, reflecting his ambition to impress the Berlin Academy and the reading public as well as the King.[5] In the text, which is preceded by an aggressively mechanistic *Vorrede* in which Kant highlighted his affinity with the ancient atomist Epicurus, while declaring himself free of his errors, and followed by an enthusiastic, *schwärmerisch* account of cosmological destructions and renewals and a discussion of the physiology and psychology of the inhabitants of other planets, he presented his nebular hypothesis. He noted at the very start of his treatise that the subject was a difficult one for two reasons. First, the derivation, by reference to lawful processes in material nature, of the existence of the celestial bodies and their arrangement in space and their motions, "the discovery of the systematic factor which ties together the great members of the created realm in the whole extent of infinity" seems to "surpass very far the forces of human reason" (*UNH* 1:221; J 81). Second, religion prefers to see the hand of God, not the forces of matter, at work in such constitutions and suspects that investigations like the present one have an atheistic intent (ibid.). In response, Kant argues that those who elevate divinely created

[4] Citations to J are to Immanuel Kant, 1755/1981, *Universal Natural History and Theory of the Heavens*, trans. Stanley L. Jaki (Edinburgh: Scottish Academic Press).

[5] Jaki 1981: 27; Zammito 2002: 55. See also Schönfeld 2000: 96–127. The 1756 printing is not, however, listed in WorldCat, and there are doubts as to whether any copies were actually sold.

nature by praising its harmonies, beauties, and purposiveness imply that "were she to be abandoned to her universal laws, she would bring forth nothing but disorder" (*UNH* 1:222–223). All is accomplished through them; there is a God, Kant concludes, "precisely because nature can proceed even in chaos in no other way than regularly and orderly" (*UNH* 1:228; J 86).

Kant assumes the matter of the world "perfectly scattered" at the outset, but with individual particles possessing different "weights" and subject to just two forces. Attraction pulls the heavier ones together into clumps, while the lighter particles exert repulsive force (Kant is thinking here of the expansion of vapors, *UNH* 1:263–264; J 114–115). Eventually, the various motions result in a state of equilibrium in which bodies disturb each other as little as possible, with planets orbiting around suns, and solar systems around their centers, and systems of systems around the center of the whole.

With an array of solar systems falling into place, individual planets undergo developmental processes, producing their meteorological and geological features, rivers, and seas. As an "active matter" materialist, albeit one who acknowledges immortal souls and perhaps a divine first cause, as distinct from a Cartesian mechanist who asserts the "passivity" of matter and God's ongoing role in causal relations, Kant sees no conflict between mechanism and "active matter." The lawfulness of these processes, which Kant refers to as "*blinde Mechanik*," is consistent with an "inherent essential striving" (*wesentliche Bestrebung*) in the primitives of nature, whereby air, water, and heat can produce such complex and novel phenomena as winds, clouds, rains, and rivers (*UNH* 1:226). The elements have "essential forces" (*wesentliche Kräfte*) to set themselves in motion and are themselves a source of life (*Quelle des Lebens*) (*UNH* 1:264; J 115). Earlier, inorganic forms are enabling of higher, later forms: thus air, water, and heat furnish the conditions needed for vegetation, which in turn is the condition of animal life.

The Epicurean philosopher Lucretius had described the Earth as first producing grasses and saplings, then birds, then as creating "wombs" in the Earth and giving birth to animals (Lucretius 1992: 441). He maintained that a younger, more fertile Earth could have engendered larger animals in the far distant past, while the old and worn-out Earth of the present could engender only small animals and insects. The view that the fertility of the Earth was native to it but wearing out was, according to Kant, widely shared, and indeed in the essay discussed below, published a year before the *Universal Natural History*, he proposed to investigate the physicochemical causes for this deterioration. Already in the *History*, however, Kant saw human reason as giving out when faced with the question of how plants and animals had come to populate the universe. He claims that the orbits of the planets, their directionality, and their elliptical shape all follow from "the simplest mechanical causes" reducible to "the easiest and clearest principles." But are we in a position to say "Give me matter, [and] I will show you how a caterpillar can be produced"? he asks (*UNH* 1:230; J 88). No, he decides: such is our "ignorance of the real inner constitution of the [living] object" that "the whole formation of the heavenly bodies, the causes of their movements, in brief, the origin of the entire present constitution of the universe will become apparent before the generation of a single herb or a caterpillar can be completely and clearly derived from mechanical causes" (*UNH* 1:230). But life just happens along, thanks to the *wesentliche Kräfte* of nature. "[M]ost planets are certainly inhabited" (*UNH* 1:354; J 186), and all are destined for habitation – the teleological purpose Kant saw in the very existence of matter, force, and motion.

Where the appearance of humans was concerned, Kant adopted a coolly naturalistic perspective. "[T]his creature seems to have been created to absorb fluids, as does a plant, and to grow, to propagate his species, and finally to age and die" (*UNH* 1:356; J 187). Most people do not develop further than is required for them to meet their elementary needs as organisms. Kant's dim view of natural man, prominent in his anthropological writings, had already made its appearance. On the planets farther from the Sun and less subject to its gravitational force, the inhabitants are composed of more tenuous stuff, less weighed down, and thus more predisposed to rationality and virtue (*UNH* 1:357–360; J 189–191).

Kant's expressed worry at this stage – probably his reason for first publishing the tract anonymously – was that he would be taken for an Epicurean atheist (*UNH* 1:222; J 82). He praises Epicurus's materialist philosophy, which "exhibited the acuteness of a great thinker," and he says that he "will not deny all agreement." The power of created matter and force testify, he claims, to the existence of the divinity, and he appeals to Descartes's widely admired earlier derivation of the world from matter and motion as offering "a more sublime idea of [God's] infinite wisdom" (*UNH* 1:227–228; J 86–87).

Although Kant seems to exempt the human soul, he takes onboard unapologetically the Epicurean thesis of the destructibility of all individual entities of whatever size, excepting only the atomic primitives of nature. As he had declared in his essay on the aging of the Earth in 1754, "All things in nature are subject to the law that the same mechanism that worked for their perfection in the beginning when they have reached that point causes their deterioration and finally leads in small steps to their destruction ... This process of nature is seen clearly in the economy of plants and animals. Exactly the same drive (*Trieb*) that causes the tree to grow brings about its death when growth is complete" (*AE* 1:198). He implicitly rejects Leibniz's eternally functioning clockwork universe and praises Newton for seeing that a mechanical universe must run down, denying that God intervenes to maintain it as Newton held.[6]

In the 1754 essay, Kant had offered four physicochemical hypotheses regarding the cause of the world's decline into senescence: desalinization, subsidence, desertification, and the exhaustion of the world spirit. He rejected the first three on lengthy observational grounds and declared the last to be the most acceptable. The world spirit, this "*wirksame Kraft*" or productive power (*AE* 1:211), is responsible for all generation and metabolism (*Oekonomie*) in the three kingdoms of nature. It does not act, Kant says, as an immaterial agent or a plastic nature, "the creations of a bold imagination," but rather as "a subtle but universally active matter [*wirksame Materie*]" like the *spiritus rector* responsible for chemical formations, "which, in the products of nature, constitutes the active principle and [as] a true Proteus is able to assume all shapes and forms." Each cycle of destruction and renewal on Earth uses some of it up, never to be restored (*AE* 1:211–212).

In the *History*, Kant considers life and death, evolution to perfection and destruction, in entire world-orders over vast time scales. The oldest worlds lie near the center of the array of systems, with the emergence and development of new worlds radiating outward, but meanwhile there is decline, repeating on a large scale the processes of destruction and renewal that we see all around us:

> Perhaps a series of millions of years and centuries will have flown by before the sphere of developed nature, in which we find ourselves, arrived at the [state of] perfection which now attends it; and perhaps there will pass an even longer period until nature makes an equally great step [back into] ... chaos. (*UNH* 1:313–314; J 154)
>
> Uncounted animals and plants become destroyed daily and are a sacrifice to transitoriness; but nature, through an inexhaustible capability of generating, brings forth no fewer [animals and plants] in other places and fills out the emptiness. Considerable pieces of the earth's surface, which we inhabit, become again buried in the sea, from which a favorable period had brought

[6] For useful discussion and translation, see Rheinhardt and Oldroyd 1982.

> them forth; but in other places nature completes the defect and brings forth other regions, which were hidden in the depths of water, in order to expand over these [regions] new riches of fruitfulness. In such a way worlds and world-orders fade away and are devoured by the abyss of eternity; however, creation is always busy in setting up new formations in other celestial regions and in repairing the loss with the gain. (*UNH* 1:317; J 157)

"Insects, flowers, men, cities, and nations are destroyed by cold, earthquakes and floods," and because "the futility which is grafted onto their finite natures" works against every world system, eternity will "usher in finally the moment of their doom through a universal decay" (*UNH* 1:317; J 157). Our world is confined "between the ruins of a collapsed and the chaos of an undeveloped nature" (*UNH* 1:319; J 159). In their old age, planets and comets crash into their suns, the suns swell up with this new nourishment, explode and redisperse their masses. Eventually the suns too collapse into a common center of numerous suns, which then blows up. It is all – though he does not use the word here – sublime. "Oh, happy is that soul when under the tumult of elements and the rubble of nature he is always set at a height from where he can see the ravages, which the frailty of the things of the world causes, roar by under his feet." At the point of our own destruction, the chains of matter will fall away from us; we experience communion with the Infinite Being and exchange the "changing scenes of nature" for rest and bliss (*UNH* 1:322; J 161).

Some sober epistemological considerations interrupt Kant's text. He notes the satisfaction for man of launching "with his imagination across the frontier of completed creation" (*UNH* 1:315; J 155) but wonders whether hypotheses can be approved simply because they are satisfying. Philosophy is "dishonoured," he says, by "free flights of fancy." Nevertheless, he declares, the present essay adduces "no other propositions than such which can truly contribute to the enlargement of our knowledge" (*UNH* 1:352; J 183). His confidence in promulgating speculative hypotheses about unobservable beginnings and endings and extraterrestrial psychology is striking, as is his conviction that the active powers of matter can produce everything in the visible world exactly as we see it.

The *History* did not make Kant's name on its first publication, though, apparently, it became a coveted book and was thrice reprinted, once in part and twice in full.[7] J. H. Lambert, for one, found Kant's theory of universal decay and cosmic renewal too harsh and Lucretian and insisted that we should confine our thoughts about development and decay to "things

[7] See below, Section 13.5.

whose birth and demise take place under our eyes" (Lambert 1761/1976: 59–63). This sentiment may have had a sobering effect on Kant. In any case, his first and second publishers fell into financial difficulties, and few copies of the *History* were actually sold.

13.2 Kant's Rejection of the Emergence of Life from Matter as Well as Hexameral Creation

In the meantime, the origins of life, design and adaptiveness, and God's purposes were topics vigorously debated in French and German letters from the midcentury onward (Philipp 1957; Schönfeld 2000: 102–103). For the materialist Maupertuis, out of the vast number of individuals produced by chance, or "blind fate" (*destin aveugle*), only a few forms possessed "order and harmony" and were able to survive and generate their kind (1750: 12). Hume, too, in the *Dialogues,* translated into German in 1781, cautiously proposed to revive the Epicurean hypothesis of the formation of the world along the same lines (1778/1998: 49). Prominent on the intelligent design side was Hermann Samuel Reimarus, whose *Abhandlung von den vornehmsten Wahrheiten der natürlichen Religion* (1755) was written to refute (according to the subtitle of his English translator), "the absurd doctrines of Epicurus, recommended by all the charms of Poetry by Lucretius, and lately retailed by several artful and admired French writers," namely, Buffon, Maupertuis, Rousseau, and La Mettrie.

Kant weighed in with *The Only Possible Argument in Support of a Demonstration of the Existence of God* in 1763, renewing his attack on physico-theology and anthropocentrism.[8] The beauty and regularity of snowflakes and flowers, their symmetry, delicacy, and proportion, should be regarded as "the necessary result of more general laws" (*OPA* 2:113), rather than as an "artificial provision," implying the operations of an intelligence. In the *Inaugural Dissertation* of 1770, Kant went on to insist that one of our rules of judgment is the Epicurean principle that "all things in the universe take place in accordance with the order of nature." He says there that all philosophers follow Epicurus, "admitting but the rarest exceptions and that only under extreme necessity" (*ID* 2:418).

However, in his 1763 essay, Kant had already retreated from the blithe emergentist account of life he appeared to favor in the *Universal Natural History*. Here, he declares unambiguously not only that the existence of

[8] Kant thought too little of human beings as products of nature to be anthropocentric. Nevertheless, the universe exists, as observed earlier, "for the sake of" the rational, intelligent beings who inhabit it.

living things "cannot be explained by appeal to the universal and necessary laws of nature," but also that "it would be absurd to regard the initial generation of a plant or animal as a mechanical effect incidentally arising from the universal laws of nature" (*OPA* 2:114). In 1781, in his essay *On the Use of Teleological Principles*, he reiterates that human reason, though it can apply itself to the theory of the Earth, cannot reach to the first production of plants and animals, which is "a science for gods," who were perhaps their creators. It can deal with the forces of nature only in existing things (*TP* 8:161–162).

In the *Critique of Judgment* of 1790, Kant decisively rejects the Epicureanism he had endorsed earlier. Epicurus, he says, "completely denies the distinction of a technic of nature from mere mechanism. Instead he adopts blind chance to explain not only ... why [nature's] products harmonize with our concept of a purpose, but even nature's mechanism ... according to the laws of motion. Hence nothing has been explained" (*CJ* 5:393). Kant could not accept that chance had any role in science. Maupertuis's and Hume's selectionism – the perishing of the randomly generated unfit – was utterly unacceptable to him.

The position Kant arrived at in 1763 and that he appears to have held thereafter is that the cause of the first members of a lineage is supernatural (later supersensible) but that in the "initial divine organization of plants and animals" there is "a capacity not merely to develop their kind thereafter in accordance with a natural law, but truly to generate their kind" (*OPA* 2:115). Once life has been installed on the planet, no further action of God is required. As Kant points out, no one supposes that when yeast reproduces itself God is involved.

13.3 Physico-theology and the *Critique of Judgment*

Kant's long-running opposition to physico-theology and its argument for the existence of God from the order and beauty of the world occasioned the juxtaposition of a book on aesthetic judgment with a book on organisms in the 1790 publication of the *Critique of Judgment.*[9] The two seemingly independent parts are connected by Kant's rejection of the physico-theological position that the beauty of living and nonliving nature is proof of its construction by a mind like ours that is well disposed toward us. So Reimarus, in his *Abhandlung* declares that:

[9] As Zuckert 2007: 3 observes, commentators tend to treat the work "piecemeal," though the parts are bound together by the concept of "purposiveness."

> the complacency which we receive from the proportion, beauty, order, and harmony of nature arises from the intelligence, design, skill, and wisdom which we observe in natural objects. Without such an observation of understanding pervading all nature, no pleasure would accrue to us but what arises from a mechanical *stimulus.* (Reimarus 1766: 161–162)

Kant agreed that our pleasure in beauty is cognitive, not simply "mechanical," and that certain "free [i.e., aesthetically captivating] forms ... seem made, as it were, for the aesthetic employment of our power of judgment" (*CJ* 5:348). However, such forms, including not only snowflakes, crystals, and stalactites, but also "the shape and color of flowers, plumage and sea-shells," can be understood in terms of nature's "ability to structure itself ... merely in accordance with chemical laws, by depositing the matter needed for this organization" (*CJ* 5:349). Although we judge them as possessing "aesthetic purposiveness," as if an idea in someone's mind guided their formation, and as if they were intended to be looked on with pleasure, the pleasure they induce does not depend on our appreciation of the divine handicraft responsible for them. Rather, the effect of "forms that are commensurate with our judgment" is a result of the way in which "their diversity and unity allow them to serve to invigorate and entertain our mental powers" (*CJ* 5:259).[10]

The second treatise of the *Critique of Judgment* suggests that "nature's ability to structure itself" extends to the formation of the bodies of plants and animals in reproduction. Kant continues to stress that the *ultimate* origination of plant and animal forms cannot have come about through mechanical laws (understood in a broad sense) operating in matter, but he allows that some mechanical process may be sufficient to bring about generation, provided some form of "organized" matter is given at the start. This comes out in his discussion of the maggot (*CJ* 5:411–412).

Maggots are generated in and from, but not by, putrefying matter. As natural scientists, Kant says, we may try to understand how matter "restructures itself" in a blind and lawful way when its elements are set free by putrefaction. Nevertheless, "we must always presuppose some original organization that itself uses mechanism, either to produce other organized forms or to develop the thing's own organized form into new shapes" (*CJ* 5:418). To explain the very possibility of maggots, we can only posit "a further principle in something [unknown] that lies beyond nature ... but that nonetheless contains the basis of nature" (*CJ* 5:412).

[10] For discussion, see Rueger 2007 and 2008.

13.4 The Antinomies of the Origins of Life

At this point one might wonder why Kant did not simply reproduce the scheme he had employed, apparently to great and long-lasting success in the *Critique of Pure Reason*, in order to reconcile universal determinism with moral freedom. In the first *Critique*, he presented as thesis and antithesis the following (paraphrased):

> Thesis: Some other form of causality than that in accord with the laws of nature must be assumed, namely the causality of freedom.
>
> Antithesis: Everything in the world takes place solely in accordance with the laws of nature. (A444/B472; A445/B473)

For Kant, the empirical perspective on human actions as causally determined in no way excluded moral responsibility (A550/B578), and in the *Critique of Practical Reason* he reinforced this claim (*CPrR* 5:99). Scientific knowledge is conditioned by our categories of matter and causality and is a way of describing the appearances. The ground of freedom may lie in the supersensible reality underlying the appearances. It is of supreme practical importance to believe in it and so mandatory to do so.

Instead of following this pattern of reconciliation, Kant presents two antinomies in the *Critique of Judgment*.[11] They are:

> The Reflective Antinomy (Judgment)
>
> Thesis: All production of material things and their forms must be judged to be possible in terms of merely mechanical laws.
>
> Antithesis: Some products of material nature cannot be judged to be possible in terms of merely mechanical laws. (*CJ* 5:387)
>
> The Determinative Antinomy (Ontology)
>
> Thesis: All production of material things is possible in terms of merely mechanical laws.
>
> Antithesis: Some production of material things is not possible in terms of merely mechanical laws. (*CJ* 5:387)

Why does he need two? Perhaps because the problem of life is not parallel to the problem of freedom. Kant was evidently persuaded that humans are like machines in their behavioral aspects; they take experiences as input, process them, and produce actions as output, and, he thinks, we might

[11] An issue raised in Breitenbach 2008.

come to a full knowledge of how these machines work. But we will never understand, he thinks, how organisms could be mechanically produced. There can never be a Newton of the grass blade (*CJ* 5:400).[12] But because he wanted to advance the determinative thesis along with the reflective antithesis, he needed two antinomies in place of one. The function of the supersensible was to make room for building forces of nature that operated in ways beyond our comprehension.

Newton had theorized an array of powers acting in matter, including electricity, magnetism, fermentation, heat, and others yet unknown, in the later editions of his *Opticks* (Newton 1730/1952: 375–376, 401), suggesting that, like attraction, they could be empirically investigated. Buffon agreed that the Cartesian conception of mechanism had to be and could legitimately be broadened, and he compared the morphological forces that act in generation to these late-Newtonian powers. These forces, he pointed out, "act upon the whole substances of bodies" and do so in constant and uniform ways. Hence they should be "ranked among mechanical principles" despite being "occult" (Buffon 1749–1757: II:61–62; 1792: II:52–53). Where generation was concerned, he proposed that "organic molecules" existing from all eternity along with inorganic molecules find their way into and also form three-dimensional molds.[13] Maupertuis advanced a related theory according to which intelligent, appetitive particles built the offspring to resemble the parent by accretion and displacement.

Kant found these two accounts untenable. In *The Only Possible Argument*, he maintained that both were "either as incomprehensible as the thing itself, or … entirely arbitrary inventions" (*OPA* 2:115). In the *Metaphysical Foundations*, he claims that hylozoism would be "the death of all natural philosophy" (*MF* 4:544). Material particles, for him, are distinctly and irrevocably dead. "[T]he possibility of living matter cannot even be thought; its concept involves a contradiction, because lifelessness, *inertia*, constitutes the essential character of matter" (*CJ* 5:394). At this point, the *Bildungstrieb* of J. F. Blumenbach appeared to him as the way forward for a theory of epigenesis, which term he understood somewhat differently from his predecessors, insofar as he required some preexisting organized body in generation rather than simply individual particles capable of joining up. As well, Blumenbach, unlike other epigenesists, appeared to respect the all-important distinction between the original formation of plants and animals and their reproduction (*CJ* 5:424).

[12] See Breitenbach 2006 and McLaughlin 2014b for discussion of this question.

[13] Roger 1997: 135–138.

Blumenbach states that in the primitive, undeveloped generative matter of organized bodies, once they are placed in a suitable environment, an active power, distinct from irritability, contractility, and sensibility, begins to operate for the entire life of the new organism, shaping it, maintaining it, and repairing it (Blumenbach 1781). Like Newton and Buffon, he insists that, while the *Trieb* is an occult quality, these must be accepted in the sciences. Its status "does not hinder us in any way whatsoever . . . from attempting to investigate the effects of this force through empirical observations and to bring them under general laws" (Blumenbach 1792: 22).[14] He even formulates some of its laws: the *Bildungstrieb* is more active in the earlier stages of development than in later stages; it acts more quickly in viviparous animals than oviparous ones; it produces less variation in brains than in other organs; and there are certain regularities in the production of monsters.

Although Blumenbach had rejected Haller's ovist theory of preformation, denying that any included miniature organism was transmitted from generation to generation, the *Bildungstrieb* was a heritable power, the necessary "original organization" in Kant's view. It thus provided for what Kant called "generic preformation" (*CJ* 5:423).[15]

For Blumenbach, despite the "immense gap between the unorganized and the organized . . . the animated and the inanimate world," even the inorganic world shows a trace of "formative power" (Blumenbach 1789/1792: 61). The *Bildungstrieb*, Kant agreed, is a species of formative, that is to say, mechanical force applying to living things. We have no idea how the formation of the caterpillar or grass blade proceeds. But we have no idea how nature builds a snowflake or a crystal or a flower, either, and they, as noted, are productions of one or more of the "merely mechanical" building forces of nature (*Bildungskraft*) (*CJ* 5:424).

Did Kant endorse the *Bildungstrieb* merely as a regulative, heuristic principle?[16] This view, I suggest, is mistaken for three reasons, as is any suggestion that Blumenbach himself adopted a regulative interpretation of his force after his reading of the *Critique of Judgment*.[17] First, if Kant had understood the *Bildungstrieb* as a regulative principle, rather than as furnishing a "constitutive" alternative to the epigenetical theories of Maupertuis and Buffon, he would have said so. Instead, he praises the

[14] On the characteristics of this force, see Lenoir 1980: 83–85.

[15] Mensch 2013 argues that this preformationist-epigenesist fusion underlies not only Kant's views about race but also his epistemology with its commitment to innate structure.

[16] Cf. Richards 2000: 19–20.

[17] Cf. Lenoir 1982/1989, which argues that Kant's regulative-constitutive distinction influentially furnished a third alternative to both *Naturphilosophie* and mechanism.

theory of epigenesis because it regards nature "as itself producing [living creatures] rather than as merely developing them; and so it minimizes appeal to the supernatural and after the first beginning leaves everything to nature" (*CJ* 5:424). Blumenbach, he declares, has properly affirmed the general application of mechanical principles: "By appealing to this principle of an original organization, a principle that is inscrutable to us, he leaves an indeterminable and yet unmistakable share to natural mechanism" (*CJ* 5:424). Heartened perhaps by this affirmation, Blumenbach emphasized the "mechanical" status of the *Trieb*, not its allegedly regulative status in the following edition of his book.[18]

Second, this "realist" position is consistent not only with the eighteenth-century Newtonian philosophy of science just summarized but with Kant's long-held belief in "*wirksame Kräfte*" as *Quelle des Lebens*, if no longer in "*wirksame Materie.*" Blumenbach's treatment of the *Bildungstrieb* with its six laws, derived from the phenomena, qualifies it as a "new mechanical principle" by eighteenth-century standards. As Kant envisioned his *Keime* and *Anlagen* as causally potent *entities* in living creatures, not as mere regulative concepts "good to think with," so, I maintain, he thought of the *Bildungstrieb* as a causally powerful, future-directed *force*. As Hume concluded that the cause of the world must possess some analogy with intelligence, implying that it needs not actually *be* intelligent, Kant argued that the only way we can conceive a morphological force that can build functional bodies is on analogy with an agent acting purposively (or purposefully, as we would say). Nevertheless, the force or forces involved may actually *be* blind. As he says in the *Opus Postumum*, "there may be quite other kinds of forces (and laws by which those forces act) than those of our thought" (*OP* 22:507). In any case, the explanation is hidden in the supersensible realm inaccessible to us.

Third, Kant was not as hostile to *Naturphilosophie* as the regulative-constitutive distinction implies. In the *Critique of Judgment,* he asks whether "nature as a whole is ... an animal" (*CJ* 5:394) with its own soul. The *anima mundi* is, he avers, a principle that must be used "with great caution," but he does not rule out its use. This remark prompted an enthusiastic letter on the topic from Salomon Maimon (11:174–176) who had just read his book; Kant does not appear to have replied. In the *Opus Postumum,* he is suddenly definite about the need to posit an underlying, indwelling, immaterial agent with a power of representation in living bodies (*OP* 22:546). Besides excitability and muscular power, there is,

[18] As Lenoir 1980: 89 himself states, his revised edition of the *Bildungstrieb* "emphasized the mechanical aspects of the model." See also Sloan 2006.

Kant surmises, a third vital power in the universe, "one all-penetrating, all-moving, etc. material, of which heat is one phenomenon. The force of organization in space and time, which contains a nonmaterial high principle, namely an effectivity according to purposes" (*OP* 22:301). And again, "Organized bodies . . . indicate an immaterial principle, and, insofar as organization extends through all parts of the world (transforming bodies and replacing dead ones with new formations in their place) indicate an *anima mundi.*" This agent, Kant supposes, "need not be a thinking being but *anima bruta*" (*OP* 22:504). It may be an "efficient cause on the analogy with an intelligence" (*OP* 22:507). The world-soul can be thought of as building its own body (*OP* 22:97), in which case, as the various organs cooperate in the living individual, "one organic whole is there for the sake of another" (*OP* 22:301).[19]

At this point, the distinction between constitutive-mechanical laws and regulative-teleological principles is no longer in evidence.[20] That distinction originated in Kant's earliest teleological projects, the reconciliation of Newtonian physics with a purposeful universe, and mechanical man with morally responsible man. It could no longer get a grip on the new teleological project of understanding how nature builds living organisms.

13.5 Teleology in Kant's Post-Critical Philosophy

Toward the end of his life, Kant found the very existence of plants and animals deeply puzzling. "Who would think," he asks in the *Opus Postumum*, "that there would be, in nature, bodies which, like works of art, are formed inwardly and outwardly, and which, furthermore, preserve their species despite the destruction of individuals, if experience were not to supply examples of such in rich measure?" (*OP* 22:357). Plants and animals seem to belong to a "separate world-system" (*OP* 22:210). "Were experience not to furnish examples of [them], the possibility of living bodies would be dismissed by everyone as fantasies of the Prince of Palagonia" (*OP* 22:383).[21]

[19] He wavers, however, on the question of whether the building force should really be considered a "soul" (*OP* 22:507).

[20] Cf. Friedman 1992c: 305. "After the execution of the *Metaphysical Foundations* . . . it becomes clear . . . that the absolute dichotomy between regulative and constitutive principles cannot be maintained."

[21] Kant refers here to the Sicilian Ferdinando Francesco Gravina Agliata, the owner of the "Villa of Monsters," a collection of grotesque statues. For his worries about the monstrous implications of transformism, see Wilson 2005.

By contrast, inanimate matter, in the form of solid bodies, is the condition of experience, so its existence is a priori (*MF* 4:481) and accordingly intelligible. Given the existence of attractive and repulsive forces as primitives, matter, with its properties of impenetrability and resistance, necessarily exists (*MF* 4:497ff.). As long as experiences and forces exist, there will be material things. But there was no way Kant could see to derive the existence of living forms from the mere possibility of sensory experience or from such minimal assumptions as two forces. True, the recognition of the basic forces (*Grundkräfte*) of "understanding" (*Verstand*) and "will" (*Wille*) in ourselves enables us to perceive the living being as like an artwork or a machine in being an active, working construction (*TP* 8:181). Organisms resemble machines insofar as they are complex entities, whose parts, including pumps, valves, vessels, and levers, seemed to be planned and assembled with care to work together. They resemble artwork in their cunning contrivance and in the beauty of flowers, feathers, and shells. Machines and artworks do not come to be either by chance or through the "mechanical" processes that Kant supposed could produce crystals and other curious mineral formations from dead matter alone. However, unlike machines, organisms have no product except their own bodies and lives and their ability to transmit existence to future generations. Unlike artworks, they were not created for human gratification.[22]

The superabundance of organisms and their seeds, and at the same time the dependence of reproduction in the human case on "opportunity ... sufficiency of food ... the moods and caprices of rulers, nay, even upon vice" (A779/B807), was difficult to square with any elevated conception of nature or human destiny. With the new emphasis on the phases of the life cycle and the science of generation, it was becoming evident that nature's purpose in producing any animal was simply to produce more of the same kind; death ensued when this task was completed, and the individual did not matter (Buffon 1749–57: XIII:j).

In the *Critique of Judgment*, Kant worried about the significance of human extinction and its implications for morality (*CJ* 5:452). "Nature," he muses in the *Opus Postumum*, "deals despotically with man. Men destroy one another like wolves ... Plants and animals overgrow and stifle one another. Nature does not observe the care and provision which they require. Wars destroy what long artifice has established and cared for" (*OP* 21:13–14; cf. 220–221). Once reproduction has occurred, nature begins to

[22] On the peculiar character of "natural purposes," see, esp., Ginsborg 2015, chapter 12.

destroy the individual (*Refl.* 1538; cf. 15:964–965).[23] In the *Conflict of the Faculties*, another late work, he alludes to the possibility that the task of morally, legally, and politically perfecting the world may be frustrated not by the aging of the Earth but by the interruption of one of the great geophysical revolutions that, according to Peter Camper and Blumenbach, had once "submerged the plant and animal kingdoms before men ever existed" (*CF* 7:160). Although it is forbidden to earthly powers to treat the human being as such, for nature, and the unknown power behind it, "the human being is but a trifle" (ibid.).

Kant's *Universal Natural History* was republished in excerpted form, minus its enthusiastic *Vorrede* and its account of exobiology and the fate of the Earth in 1791, together with an account of cosmogenesis by William Herschel. Then, despite Kant's expressed desire not to include his works written before 1770 in any collection of his writings (12:208), the full text of the *History* was printed in 1797.

This republication event should surprise and perhaps astonish us. We may reckon in this connection with two considerations. First, Kant was not, by his own admission, a well man in 1797.[24] For decades, he had suffered from headaches and visual disturbances suggestive of a brain tumor or insidious neurodegenerative condition. He had withdrawn from teaching in 1794, and medical historians agree that he began to exhibit severe confusion and ataxia in 1796.[25] He had published a strange essay on the weather in 1794, which, despite "endless references to critical reasoning," was back, as Jaki comments, in the old speculative register of the *History* (1981: 35). Notwithstanding its philosophical creativity and profundity, the repetitious phraseology of the *Critique of Judgment* of 1790 and the overall disorganization of that work by comparison with the relative clarity of Kant's earlier treatments of the problem of life might be seen as foreshadowing the oncoming health crisis.

Second, Kant's visionary, cosmological perspective and his ambivalence over the aging of the Earth had long been elements of his thinking, sometimes expressing themselves in a floridly speculative manner. On

[23] Quoted by Mary J. Gregor in the introduction to her translation of Kant's *The Conflict of the Faculties* (see Gregor 1979: xxiii).

[24] Kant described his latest illness, which he supposed incurable, in the conclusion to the *Conflict of the Faculties* (1798) as "a kind of gout that has to some extent penetrated my brain," causing him to feel disoriented, weakened, and dulled (*CF* 7:112–113).

[25] Cf. Marchand 1997; Miranda et al. 2010; Fellin and Bley 1997. Kant later developed problems with gait, sense of smell (a symptom of central nervous system degeneration), nightmares, hallucinations, and obsessions.

one hand, the cycles of birth, maturation, and death to which the worlds and world-orders were fated according to science held out the promise of ever greater moral and intellectual achievements elsewhere in the evolving universe. On the other hand, they threatened existence in the here and now with moral and political futility, necessitating a metaphysical response that Kant found in his noumenal world inaccessible to science. The "Critical" works are imbued with elements of what Carl du Prel in 1889 referred to as "Kant's mystical *Weltanschauung*," which returned to clear visibility in the "post-Critical" and malady-infused *Opus Postumum.*

To conclude, although he believed that living things maintained themselves and generated their progeny by natural mechanisms without supernatural assistance, Kant doubted from the beginning of his philosophical career that the kind of causality theorized in the physical sciences would ever be able to explain the origins of life on the planet. He was tempted nevertheless to believe, and thought it morally essential to believe, that certain forces tended toward development and improvement, not only in the embryo but in living nature generally, that the late appearance of rational beings in the cosmos was not accidental, and that the mysterious productive forces of nature might bring forth a superior race to succeed us.

Bibliography

Adams, Robert. 1994. *Leibniz: Determinist, Theist, Idealist.* Oxford University Press.

Allison, Henry E. 1996. *Idealism and Freedom.* Cambridge University Press.

2001. *Kant's Theory of Taste.* Cambridge University Press.

Ameriks, Karl. 1978. "Kant's Transcendental Deduction as a Regressive Argument." *Kant-Studien* 69: 273–285.

2000. *Kant and the Fate of Autonomy.* Cambridge University Press.

2006. *Kant and the Historical Turn.* Oxford University Press.

2012. *Kant's Elliptical Path.* Oxford University Press.

2016. "On the Many Senses of 'Self-Determination,'" in K. Moran and J. von Platz (eds.), *Freedom and Spontaneity in Kant.* Cambridge University Press.

Armstrong, David. 1983. *What Is a Law of Nature?.* Cambridge University Press.

Beatty, John. 1995. "The Evolutionary Contingency Thesis," in G. Wolters and J. G. Lennox (eds.), *Concepts, Theories and Rationality in Biological Science.* University of Pittsburgh Press.

Beck, Lewis White. 1967. "Once More unto the Breach: Kant's Answer to Hume, Again." *Ratio* 9: 33–37.

Beiser, Frederick. 2006. "Kant and *Naturphilosophie,*" in M. Friedman and A. Nordmann (eds.), *The Kantian Legacy in Nineteenth Century Science.* MIT Press. Pp. 7–26.

Bennett, Jonathan. 1966. *Kant's Analytic.* Cambridge University Press.

Bertoloni Meli, Domenico. 1993. *Equivalence and Priority: Newton versus Leibniz.* Oxford University Press.

Bird, Alexander. 2007. *Nature's Metaphysics: Laws and Properties.* Oxford University Press.

Blumenbach, Johann Friedrich. 1781/1789. *Ueber den Bildungstrieb.* Translated by A. Crichton. T. Cadell, 1792.

Brandt, Reinhard. 1989. "The Deductions in the *Critique of Judgment*: Comments on Hampshire and Horstmann," in E. Förster (ed.), *Kant's Transcendental Deductions.* Stanford University Press.

Breitenbach, Angela. 2006. "Mechanical Explanation of Nature and Its Limits in Kant's *Critique of Judgment.*" *Studies in History and Philosophy of Biological and Biomedical Science* 37: 694–711.

2008. "Two Views on Nature: A Solution to Kant's Antinomy of Mechanism and Teleology." *British Journal for the History of Philosophy* 16: 351–369.

2009a. *Die Analogie von Vernunft und Natur: Eine Umweltphilosophie nach Kant*. Walter de Gruyter.

2009b. "Teleology in Biology: A Kantian Approach." *Kant Yearbook* 1: 31–56.

2014. "Biological Purposiveness and Analogical Reflection," in I. Goy and E. Watkins (eds.), *Kant's Theory of Biology*. Walter de Gruyter.

Forthcoming. "Laws and Ideal Unity," in W. Ott and L. Patton (eds.), *Laws of Nature*. Oxford University Press.

Brittan, Gordon. 1978. *Kant's Theory of Science*. Princeton University Press.

Buchdahl, Gerd. 1965. "Causality, Causal Law, and Scientific Theory in the Philosophy of Kant." *British Journal for the Philosophy of Science* 16: 187–208.

Metaphysics and the Philosophy of Science. Blackwell.

Buchwald, J. Z. 1994. *The Creation of Scientific Effects: Heinrich Hertz and Electric Waves*. University of Chicago Press.

Buffon, George-Louis Leclerc. 1749–1804. *Histoire Naturelle, générale et particulière, avec la description du Cabinet du Roy*, 46 vols. Imprimerie du Roi.

1792. *Buffon's Natural History Containing a Theory of the Earth, a General History of Man, of the Brute Creation, and of Vegetables, Minerals, &c.*, 10 vols. Translated by J. Smith Barr. J. S. Barr.

Chignell, Andrew. 2014. "Modal Motivations for Noumenal Ignorance: Knowledge, Cognition, and Coherence." *Kant-Studien* 105: 573–597.

Choi, Sungho and Fara, Michael. 2016. "Dispositions", The Stanford Encyclopedia of Philosophy (Spring 2016 Edition), Edward N. Zalta (ed.), URL = <https://plato.stanford.edu/archives/spr2016/entries/dispositions/>.

Clarke, Ellen. 2010. "The Problem of Biological Individuality." *Biological Theory* 5: 312–325.

2013. "The Multiple Realizability of Biological Individuals." *Journal of Philosophy* 60: 413–435.

Cohen, Alix. Forthcoming. "Kant on Evolution: A Re-Evaluation," in J. Callanan and L. Allais (eds.), *Kant on Animals*. Oxford University Press.

Cohen, I. Bernard 1980. *The Newtonian Revolution*. Cambridge University Press.

Cohen, I. Bernard and Anne Whitman (eds.). 1999. *The Principia: Mathematical Principles of Natural Philosophy*. University of California Press.

Correia, Fabrice 2011. "On the Reduction of Necessity to Essence." *Philosophy and Phenomenological Research* 84: 639–653.

Darrigol, Olivier. 2000. *Electrodynamics from Ampère to Einstein*. Oxford University Press.

Descartes, René. 1985. *The Philosophical Writings of Descartes*, 2 vols. Edited and translated by J. Cottingham, R. Stoothoff, and D. Murdoch. Cambridge University Press.

Dretske, Fred. 1977. "Laws of Nature." *Philosophy of Science* 44: 248–268.

Du Prel, Carl. 1889. *Kant's mystische Weltanschauung, in Immanuel Kant's Vorlesungen ueber die Psychologie*. Ernst Guenther.

Dyck, Corey. 2014. *Kant and Rational Psychology*. Oxford University Press.

Edwards, Jeffrey. 2000. *Substance, Force, and the Possibility of Knowledge: On Kant's Philosophy of Material Nature*. University of California Press.

Ellis, Brian. 2001. *Scientific Essentialism*. Cambridge University Press.

Ereshefsky, Marc. 2001. *The Poverty of the Linnaean Hierarchy: A Philosophical Study of Biological Taxonomy*. Cambridge University Press.

Fellin, Renato, and Alessandro Blè. 1997. "The Disease of Immanuel Kant." *The Lancet* 350: 1771–1773.

Fine, Kit. 1994. "Essence and Modality," in J. Tomberlin (ed.), *Philosophical Perspectives 8: Logic and Language*. Ridgeview. Pp. 1–16.

2002. "Varieties of Necessity," in T. Szabó Gendler and J. Hawthorne (eds.), *Conceivability and Possibility*. Oxford University Press. Pp. 253–281.

2012. "Guide to Ground," in F. Correia and B. Schnieder (eds.), *Metaphysical Grounding: Understanding the Structure of Reality*. Cambridge University Press. Pp. 37–81.

Floyd, Juliet. 1998. "Heautonomy: Kant on Reflective Judgment and Systematicity," in H. Parret (ed.), *Kants Ästhetik/Kant's Aesthetics/L'esthétique De Kant*. Walter de Gruyter. Pp. 192–218.

Förster, Eckart. 2000. H. Parret (ed.), *Kant's Final Synthesis: An Essay on the Opus postumum*. Harvard University Press.

Friedman, Michael. 1983. *Foundations of Space-Time Theories: Relativistic Physics and Philosophy of Science*. Princeton University Press.

1986. "Metaphysical Foundations of Newtonian Science," in R. E. Butts (ed.), *Kant's Philosophy of Physical Science*. Reidel. Pp. 25–60.

1992a. *Kant and the Exact Sciences*. Harvard University Press.

1992b. "Causal Laws and the Foundations of Natural Science," in P. Guyer (ed.), *The Cambridge Companion to Kant*. Cambridge University Press. Pp. 161–199.

1992c. "Regulative and Constitutive." *The Southern Journal of Philosophy* 30: 73–102.

2001. *Dynamics of Reason: The 1999 Kant Lectures at Stanford University*. Center for the Study of Language and Information Publications.

2008. "Einstein, Kant, and the a Priori," in M. Massimi (ed.), *Kant and Philosophy of Science Today, Royal Institute of Philosophy Supplement: 63*. Cambridge University Press. Pp. 95–112.

2010. "Synthetic History Reconsidered," in M. Domski and M. Dickson (eds.), *Discourse on a New Method: Reinvigorating the Marriage of History and Philosophy of Science*. Open Court. Pp. 571–813.

2012a. "Reconsidering the Dynamics of Reason." *Studies in History and Philosophy of Science* 43: 47–53.

2012b. "The *Prolegomena* and Natural Science," in H. Lyre and O. Schliemann (eds.), *Kant's Prolegomena. Ein kooperativer Kommentar*. Klostermann. Pp. 299–326.

2012c. "Newton and Kant: Quantity of Matter in the *Metaphysical Foundations of Natural Science*." *The Southern Journal of Philosophy* 50: 482–503.

2013. *Kant's Construction of Nature: A Reading of the Metaphysical Foundations of Natural Science*. Cambridge University Press.

2014. "Laws of Nature and Causal Necessity." *Kant-Studien* 105: 531–553.

Geiger, Ido. 2003. "Is the Assumption of a Systematic Whole of Empirical Concepts a Necessary Condition of Knowledge?" *Kant-Studien* 94: 273–298.

Geroch, Robert. 1978. *General Relativity from A to B.* University of Chicago Press.

Ginsborg, Hannah. 1992. "Kant and the Systematicity and Purposiveness of Nature." Unpublished ms.

1998. "Korsgaard on Choosing Nonmoral Ends." *Ethics* 109: 5–21.

2015. *The Normativity of Nature: Essays on Kant's Critique of Judgement.* Oxford University Press.

Gregor, Mary J. 1979. "Introduction," in Immanuel Kant, *Conflict of the Faculties.* Translated by M. J. Gregor. Abaris.

Grier, Michelle. 1997. "Kant on the Illusion of a Systematic Unity of Knowledge." *History of Philosophy Quarterly* 14: 1–28.

2001. *Kant's Doctrine of Transcendental Illusion.* Cambridge University Press.

Gurtin, Morton, Eliot Fried, and Lallit Anand. 2010. *The Mechanics and Thermodynamics of Continua.* Cambridge University Press.

Guyer, Paul. 1987. *Kant and the Claims of Knowledge.* Cambridge University Press.

1990. "Reason and Reflective Judgment: Kant on the Significance of Systematicity." *Noûs* 24: 17–43.

1997. *Kant and the Claims of Taste*, second edition (first edition published in 1979). Cambridge University Press.

2000. "The Unity of Nature and Freedom: Kant's Conception of the System of Philosophy," in S. S. Sedgwick (ed.), *The Reception of Kant's Critical Philosophy.* Cambridge University Press. Pp. 19–63. Reprinted in Guyer 2005.

2003a. "Kant's Answer to Hume." *Philosophical Topics* 31: 127–164. Reprinted in Guyer 2008.

2003b. "Kant on the Systematicity of Nature: Two Puzzles." *History of Philosophy Quarterly* 20: 277–295.

2005. *Kant's System of Nature and Freedom.* Oxford University Press.

2008. *Knowledge, Reason, and Taste: Kant's Response to Hume.* Princeton University Press. Pp. 71–123.

Hamel, Georg. 1909. "Über die Grundlagen der Mechanik." *Mathematische Annalen* 66: 350–397.

Hebbeler, James. 2015. "Kant on Necessity, Insight, and a Priori Knowledge." *Archiv für Geschichte der Philosophie* 97: 34–65.

Herschel, William. 1791. *Über den Bau des Himmels: Drey Abhandlungen aus dem Englischen übersetzt. Nebst einem authentischen Auszug aus Kants allgemeiner Naturgeschichte und Theorie des Himmels.* Friedrich Nicolovius.

Hume, David. 1739–1740/2007. *A Treatise of Human Nature.* Edited by D. Fate and M. J. Norton. 2 vols. Clarendon Press.

1748/2000. *An Enquiry Concerning Human Understanding.* Edited by T. L. Beauchamp. Clarendon Press.

1755. *Vermischte Schriften*. Translated by G. C. Grund and A. H. Holle. Hamburg and Leipzig.

1778/1998. *Dialogues Concerning Natural Religion*. Edited by R. H. Popkin. Hackett.

1790–1792. *Über die menschliche Natur*. Translated by L. H. Jakob, 3 vols. Hemmerde and Schwetschke.

Irving, J. H., and J. G. Kirkwood. 1950. "The Statistical Mechanical Theory of Transport Processes IV. The Equations of Hydrodynamics." *The Journal of Chemical Physics* 18: 817–829.

Jaki, Stanley L. 1981. "Introduction," in his (tr.) I. Kant 1755. *Universal Natural History and Theory of the Heavens*. Scottish Academic Press.

Joos, Georg 1934. *Theoretical Physics*. Translated by I. M. Freeman. Hafner.

Kant, Immanuel. 1900– . *Gesammelte Schriften*, 29 vols. Edited by Deutsche (formerly Königlich Preussische) Akademie der Wissenschaften. De Gruyter.

1992a. *Lectures on Logic*. Translated and edited by J. M. Young. Cambridge University Press.

1992b. *Theoretical Philosophy 1755–1770*. Translated and edited by D. Walford and R. Meerbote. Cambridge University Press.

1995. *Opus Postumum*. Translated and edited by E. Förster and M. Rosen. Cambridge University Press.

1996. *Practical Philosophy*. Translated and edited by M. J. Gregor. Cambridge University Press.

1997. *Lectures on Metaphysics*. Translated and edited by K. Ameriks and S. Naragon. Cambridge University Press.

1998. *Critique of Pure Reason*. Translated and edited by P. Guyer and A. W. Wood. Cambridge University Press.

2001. *Critique of the Power of Judgment*. Translated and edited by P. Guyer and E. Matthews. Cambridge University Press.

2002. *Theoretical Philosophy after 1781*. Edited by H. E. Allison and P. Heath. Translated by G. Hatfield, M. Friedman, H. E. Allison and P. Heath. Cambridge University Press.

2004. *Metaphysical Foundations of Natural Science*. Translated and edited by M. Friedman. Cambridge University Press.

2012. *Natural Science*. Edited by E. Watkins. Translated by L. W. Beck, J. B. Edwards, O. Reinhardt, M. Schönfeld, and E. Watkins. Cambridge University Press.

Kemp-Smith, Norman. 1992. *A Commentary to Kant's Critique of Pure Reason*. Macmillan.

Kennefick, Daniel. 2009. "Testing Relativity from the 1919 Eclipse: A Question of Bias." *Physics Today* 62: 37–42.

Kitcher, Philip. 1986. "Projecting the Order of Nature," in R. E. Butts (ed.), *Kant's Philosophy of Physical Science*. Reidel. Pp. 201–235.

1994. "The Unity of Science and the Unity of Nature," in Paolo Parrini (ed.), *Kant and Contemporary Epistemology*. Kluwer. Pp. 253–272.

1996. "Aprioristic Yearnings." *Erkenntnis* 44: 397–416.

Kleingeld, Pauline. 1998. "The Conative Character of Reason in Kant's Philosophy." *The Journal of the History of Philosophy* 36: 77–97.

Kreines, James. 2009. "Kant on the Laws of Nature and the Limitation of Our Knowledge." *European Journal of Philosophy* 17: 527–558.

Lagrange, Joseph-Louis. 1777. "Remarques générales sur le mouvement des plusieurs corps qui s'attirent mutuellement en raison inverse des carrés des distances," in *Nouveaux Mémoires de l'Académie royale des sciences et belles-lettres*. Berlin. Pp. 155–172.

Lambert, J. H. 1761/1976. *Cosmological Letters on the Arrangement of the World Edifice*. Edited and translated by S. L. Jaki. Scottish Academic Press.

Langton, Rae. 1998. *Kantian Humility and Things in Themselves*. Oxford University Press.

Leibniz, Gottfried. 1989. *Philosophical Essays*. Edited and translated by R. Ariew and D. Garber. Hackett.

Lenoir, Timothy. 1980. "Kant, Blumenbach and Vital Materialism in German Biology." *Isis* 71: 77–108.

1982/1989. *The Strategy of Life*. University of Chicago.

Lewis, David. 1997. "Finkish Dispositions." *The Philosophical Quarterly* 47: 143–158.

Longuenesse, Béatrice. 1998. *Kant and the Capacity to Judge: Sensibility and Discursivity in the Transcendental Analytic of the Critique of Pure Reason*. Princeton University Press.

Lucretius, Titus Carus. 1992. *De Rerum Natura*. Translated by W. H. Rouse. Revised by M. F. Smith. Harvard University Press.

Malvern, Lawrence. 1969. *Introduction to the Mechanics of a Continuous Medium*. Prentice-Hall.

Marchand, J. C. 1997. "Was Emmanuel Kant's Dementia Symptomatic of a Frontal Tumor?" *Revue Neurologique* 153: 35–39.

Massimi, Michela. (ed.) 2008. *Kant and Philosophy of Science Today*, Royal Institute of Philosophy Supplement: 63. Cambridge University Press.

2011. "Kant's Dynamical Theory of Matter in 1755, and Its Debt to Speculative Newtonian Experimentalism." *Studies in History and Philosophy of Science* 42: 525–543.

2014. "Prescribing Laws to Nature. Part I. Newton, the Pre-Critical Kant, and Three Problems about the Lawfulness of Nature." *Kant-Studien* 105: 491–508.

Forthcoming. "Why Kant Was Not a Projectivist about Laws," in *Proceedings of the 12th Kant Congress*.

Maupertuis, Pierre Louis Moreau de [aka Dr. Baumann]. 1750. *Essay de cosmologie*. Paris.

[aka Dr. Baumann]. 1754. *Essai sur la formation de corps organisés*. Paris and Berlin.

McKinsey, John C. C. and Patrick Suppes. 1955. "On the Notion of Invariance in Classical Mechanics." *British Journal for the Philosophy of Science* 5: 290–302.

McLaughlin, Peter. 1990. *Kant's Critique of Teleology in Biological Explanation: Antinomy and Teleology*. Mellon Press.

2003. "Newtonian Biology and Kant's Mechanistic Concept of Causality," in P. Guyer (ed.), *Kant's Critique of the Power of Judgment*. Rowman and Littlefield. Pp. 209–218.

2014a. "Transcendental Presuppositions and Ideas of Reason." *Kant-Studien* 105: 554–572.

2014b. "Mechanical Explanation in the 'Critique of the Teleological Power of Judgment,'" in Ina Goy and Eric Watkins (eds.), *Kant's Theory of Biology*. De Gruyter. Pp. 149–166.

McNulty, Bennett. 2015. "Rehabilitating the Regulative Use of Reason: Kant on Empirical and Chemical Laws." *Studies in History and Philosophy of Science* 54: 1–10.

Mensch, Jennifer. 2013. *Kant's Organicism*. University of Chicago Press.

Miranda, Marcelo, Andrea Slachevsky, and Diego Garcia-Borreguero. 2010. "Did Immanuel Kant Have Dementia with Lewy Bodies and REM Behavior Disorder?" *Sleep Medicine* 11: 586–588.

Misner, C. W., K. S. Thorne, and J. A. Wheeler. 1973. *Gravitation*. Freeman

Mitchell, Sandra D. 2000. "Dimensions of Scientific Law." *Philosophy of Science* 67: 242–265.

Mumford, Stephen. 2004. *Laws in Nature*. Routledge.

Murdoch, A. Ian. 2012. *Physical Foundations of Continuum Mechanics*. Cambridge University Press.

Nicholson, Daniel J. 2014. "The Return of the Organism as a Fundamental Explanatory Concept in Biology." *Philosophy Compass* 9: 347–359.

Okasha, Samir. 2002. "Darwinian Metaphysics: Species and Question of Essentialism." *Synthese* 131: 191–213.

O'Shea, James. 1997. "The Needs of Understanding: Kant on Empirical Laws and Regulative Ideals." *International Journal of Philosophical Studies* 5: 216–254.

Ott, Walter. 2009. *Causation and Laws of Nature in Early Modern Philosophy*. Oxford University Press.

Philipp, Wolfgang. 1957. *Das Werden der Aufklärung in theologiegeschichtlicher Sicht*. Vanderhoeck and Ruprecht.

Pollok, Konstantin. 2001. Kants *Metaphysische Anfangsgründe der Naturwissenschaft*. Meiner Verlag.

2014. "'The Understanding Prescribes Laws to Nature': Spontaneity, Legislation, and Kant's Transcendental Hylomorphism." *Kant-Studien* 105: 509–530.

Prior, E., R. Pargetter, and F. Jackson. 1982. "Three Theses about Dispositions." *American Philosophical Quarterly* 19: 251–257.

Quarfood, Marcel. 2004. *Transcendental Idealism and the Organism: Essays on Kant*. Almqvist & Wiksell.

Rauscher, Frederick. 2010. "The Appendix to the Transcendental Dialectic and the Canon of Pure Reason: The Positive Role of Reason," in P. Guyer (ed.), *The Cambridge Companion to Kant's Critique of Pure Reason*. Cambridge University Press.

Reath, Andrews. 2006. *Agency and Autonomy in Kant's Moral Theory*. Oxford University Press.

Reichenbach, Hans. 1920. *Relativitätstheorie und Erkenntnis Apriori*. Springer

Reimarus, Hermann Samuel. 1766. *The Principal Truths of Natural Religion Defended and Illustrated, in Nine Dissertations: Wherein the Objections of Lucretius, Buffon, Maupertius, Rousseau, La Mettrie, and Other Ancient and Modern Followers of Epicurus Are Considered and Their Doctrines Refuted.* Translated by R. Wynne B. Law.

Reinhardt, O., and D. R. Oldroyd. 1982 "Kant's Thoughts on the Aging of the Earth." *Annals of Science* 39: 349–369.

Resnick, Robert. 1968. *Introduction to Special Relativity*. Wiley.

Richards, Robert. 2000. "Kant and Blumenbach on the *Bildungstrieb*: A Historical Misunderstanding." *Studies in History and Philosophy of Biological and Biomedical Sciences* 31: 11–32.

Ripstein, Arthur. 2009. *Force and Freedom: Kant's Legal and Political Philosophy*. Harvard University Press.

Roger, Jacques. 1997. *Buffon: A Life in Natural History*. Translated by S. Lucille Bonnefoi. Cornell University Press.

Rohlf, Michael. 2014. "The Rationality of Induction in Kant (and Hume)." *Idealistic Studies* 43: 153–169.

Rosen, Gideon. 2010. "Metaphysical Dependence: Grounding and Reduction," in B. Hale and A. Hoffmann (eds.), *Modality: Metaphysics, Logic and Epistemology*. Oxford University Press. Pp. 109–136.

Rosenberg, Alexander. 1994. *Instrumental Biology or the Disunity of Science*. University of Chicago Press.

Rudwick, Martin J. 2014. *Earth's Deep History*. University of Chicago Press.

Rueger, Alexander. 2007. "Kant and the Aesthetics of Nature." *British Journal of Aesthetics* 47: 138–155.

2008. "Beautiful Surfaces: Kant on Free and Adherent Beauty in Nature and Art." *British Journal for the History of Philosophy* 16: 535–557.

Rush, Fred L. Jr. 2000. "Reason and Regulation in Kant." *The Review of Metaphysics* 53: 837–862.

Rutherford, Donald. 1993. "Nature, Laws, and Miracles: The Roots of Leibniz's Critique of Occasionalism," in S. Nadler (ed.), *Causation in Early Modern Philosophy*. Pennsylvania State University Press.

Schönfeld, Martin 2000. *The Philosophy of the Young Kant: The Pre-Critical Project*. Oxford University Press

Schopenhauer, Arthur. 1813/2012. *On the Fourfold Root of the Principle of Sufficient Reason*, second edition. Translated by D. E. Cartwright, E. E. Erdmann, and C. Janaway. Cambridge University Press.

Sloan, Philip R. 2006. "Kant on the History of Nature: The Ambiguous Heritage of the Critical Philosophy for Natural History." *Studies in History and Philosophy of Biological and Biomedical Sciences* 37: 637–648.

Smith, George E. 2002. "The Methodology of the *Principia*," in I. B. Cohen and G. E. Smith (eds.), *The Cambridge Companion to Newton*. Cambridge University Press. Pp. 138–173.

2012. "How Newton's *Principia* Changed Physics," in A. Janiak and E. Schliesser (eds.), *Interpreting Newton: Critical Essays*. Cambridge University Press. Pp. 360–395.

2014. "Closing the Loop: Testing Newtonian Gravity, Then and Now," in Z. Biener and E. Schliesser (eds.), *Newton and Empiricism*. Oxford University Press. Pp. 252–351.

Smith, Sheldon. 2013. "Kant's Picture of Monads in the *Physical Monadology*." *Studies in History and Philosophy of Science* 44: 102–111.

Sober, Elliott. 1997. "Two Outbreaks of Lawlessness in Recent Philosophy of Biology." *Philosophy of Science* 64, Supplement: 458–467.

Stan, Marius. 2014. "Unity for Kant's Natural Philosophy." *Philosophy of Science* 81: 423–443.

Stark, Werner (ed.). 2004. *Immanuel Kant, Vorlesung zur Moralphilosophie*. Walter de Gruyter.

Stein, Howard. 1970. "On the Notion of Field in Newton, Maxwell, and Beyond," in R. Stuewer (ed.), *Historical and Philosophical Perspective of Science*. University of Minnesota Press. Pp. 264–287.

Sterelny, Kim. 2001. "The Reality of Ecological Assemblages: A Palaeoecological Puzzle." *Biology and Philosophy* 16: 437–461.

Sutherland, Daniel. 2014. "Kant on the Construction and Composition of Motion in the Phoronomy." *Canadian Journal of Philosophy* 44: 686–718.

Teufel, Thomas. 2012. "What Does Kant Mean by 'Power of Judgment' in His *Critique of the Power of Judgment*?" *Kantian Review* 17: 297–326.

Tolley, Clinton. 2006. "Kant and the Nature of Logical Laws." *Philosophical Topics* 34: 371–407.

Tooley, Michael. 1977. "The Nature of Laws." *Canadian Journal of Philosophy* 74: 667–698.

Truesdell, Clifford. 1991. *A First Course in Rational Continuum Mechanics*, second edition. Boston: Academic Press.

Van Fraassen, Bas. 1989. *Laws and Symmetries*. Oxford University Press.

Vetter, Barbara. 2012. "Dispositional Essentialism and the Laws of Nature," in A. Bird, B. Ellis, and H. Sankey (eds.), *Properties, Powers, and Structures: Issues in the Metaphysics of Realism*. Routledge.

Walsh, Denis. 2015. *Organisms, Agency, and Evolution*. Cambridge University Press.

Warren, Daniel. 2001. *Reality and Impenetrability in Kant's Philosophy of Nature*. Routledge.

Watkins, Eric. 1997. "The Laws of Motion from Newton to Kant." *Perspectives on Science* 5: 311–348.

1998a. "Kant's Justification of the Laws of Mechanics." *Studies in History and Philosophy of Science* 29: 539–560.

1998b. "The Argumentative Structure of Kant's *Metaphysical Foundations of Natural Science*." *Journal of the History of Philosophy* 36: 567–593.

2000. "Kant on Rational Cosmology," in E. Watkins (ed.), *Kant and the Sciences*. Oxford University Press. Pp. 70–89.

2005. *Kant and the Metaphysics of Causality*. Cambridge University Press.

2009. "The Antinomy of Teleological Judgment." *Kantian Yearbook* 1: 197–221.

2013. "Kant on *Infima Species*," in Claudio La Rocca (ed.), *Kant und die Philosophie in weltbürgerlicher Absicht*. De Gruyter. Vol. 5, pp. 283–294.

2014a. "What Is, for Kant, a Law of Nature?" *Kant-Studien* 105: 471–490.

2014b. "Nature in General as a System of Ends," in I. Goy and E. Watkins, *Kant's Theory of Biology*. Walter de Gruyter. Pp. 117–130.

Wilson, Robert A., and Matthew Barker. 2017. "The Biological Notion of Individual," *The Stanford Encyclopedia of Philosophy*. (Spring 2017 Edition), Edward N. Zalta (ed.), forthcoming URL = <https://plato.stanford.edu/archives/spr2017/entries/biology-individual/>.

Wilson, Catherine. 2005. "Kant and the Speculative Sciences of Origins," in J. E. H. Smith (ed.), *The Problem of Animal Generation in the 17th and 18th Centuries*. Cambridge University Press. Pp. 375–401.

Wilson, Mark. 2013. "What Is 'Classical Mechanics' Anyway?," in R. Batterman (ed.), *The Oxford Handbook of Philosophy of Physics*. Oxford University Press. Pp. 43–106.

Zammito, John. 1992. *The Genesis of Kant's Critique of Judgment*. University of Chicago Press.

2012. "The Lenoir Thesis Revisited: Blumenbach and Kant." *Studies in History and Philosophy of Biological and Biomedical Sciences* 43: 120–132.

Zuckert, Rachel. 2007. *Kant on Beauty and Biology: An Interpretation of the Critique of Judgment*. Cambridge University Press.

Index

Lightning Source UK Ltd.
Milton Keynes UK
UKOW06n0422211217
314804UK00011B/138/P